Cognitive-Affective Neuroscience of Depression and Anxiety Disorders

Cognitive-Affective Neuroscience of Depression and Anxiety Disorders

Dan J Stein, MD, PhD
Director, MRC Unit on Anxiety Disorders, Department of Psychiatry, University of Stellenbosch, Cape Town, South Africa

and

University of Florida, Gainesville, USA

LONDON AND NEW YORK

First published in the United Kingdom in 2003
by Martin Dunitz Ltd, Taylor & Francis Group plc, 11 New Fetter Lane, London
EC4P 4EE

Tel.: +44 (0) 20 7482 2202
Fax.: +44 (0) 20 7267 0159
E-mail: info.dunitz@tandf.co.uk
Website: http://www.dunitz.co.uk

A CIP record for this book is available from the British Library.

ISBN 1-84184-100-5

Distributed in the USA by
Fulfilment Center
Taylor & Francis
7625 Empire Drive
Florence, KY 41042, USA
Toll Free Tel.: +1 800 634 7064
E-mail: cserve@routledge_ny.com

Distributed in Canada by
Taylor & Francis
74 Rolark Drive
Scarborough, Ontario M1R 4G2, Canada
Toll Free Tel.: +1 877 226 2237
E-mail: tal_fran@istar.ca

Distributed in the rest of the world by
ITPS Limited
Cheriton House
North Way
Andover, Hampshire SP10 5BE, UK
Tel.:+44 (0)1264 332424
E-mail: reception@itps.co.uk

An abridged version of this volume is published by Martin Dunitz under the title *Serotonergic Neurocircuitary of Mood and Anxiety Disorders* (ISBN 1–84184–303–2)

Composition by Wearset Ltd, Boldon, Tyne and Wear

Printed and bound in Italy by Printer Trento S.r.l.

Contents

Acknowledgements

The support of the Medical Research Council of South Africa is gratefully acknowledged. Permission of the University of Stellenbosch to reprint the illustrations is appreciated, with particular thanks to Carol Lochner for her artistic collaboration. While a range of primary sources were used to develop the illustrations, the University of Washington Digital Anatomist Program and Salloway and Colleagues' volume on the Neuropsychiatry of Limbic and Subcortical Disorders were particularly helpful. Professor David Nutt provided valuable feedback on an early draft.

Heather, Gabriella, Joshua, and our new arrival, consistently provided encouragement and inspiration.

Preface

This volume is to some extent about the diagnosis, evaluation, and treatment of the mood and anxiety disorders mostly commonly seen in general clinical practice. Chapters are devoted to major depression, generalized anxiety disorder, obsessive-compulsive disorder, panic disorder, post-traumatic stress disorder, and social anxiety disorder.

This volume is to a greater extent about the psychobiology of these conditions. Each chapter highlights the mediating neurocircuitry, and relates this to the relevant cognitive-affective functions, proximal factors (e.g. neurotransmitter systems), and distal factors (e.g. the evolutionary basis) involved in the disorder under discussion.

The volume is, however, ultimately an attempt to develop an integrated conceptual approach to depression and the anxiety disorders. While it is important for clinicians to keep up to date with clinical issues and with new data, what excites many psychiatrists is the question of how best to think about the brain–mind and its psychopathology.

The approach taken here to the brain–mind and to its psychopathology draws on current developments in cognitive-affective neuroscience and evolutionary theory. I hope that this will not only provide practicing clinicians with a novel and thought-provoking perspective on their patients, but also comprise a useful framework for thinking about appropriate management.

Dan J Stein, MD, PhD

chapter 1

Introduction

Why this volume?

This volume is written for clinicians who work with depression and the anxiety disorders. Chapters cover major depression, as well as the most important anxiety disorders seen in clinical practice – generalized anxiety disorder, obsessive-compulsive disorder, panic disorder, post-traumatic stress disorder, and social anxiety disorder (social phobia). The contents include clinically relevant material such as diagnosis, evaluation, and treatment.

The volume aims to provide an integrative conceptual approach to depression and anxiety disorders. Although a range of important advances has taken place in understanding these conditions, important clinical and conceptual problems remain. The approach here draws on developments in cognitive-affective neuroscience, and emphasizes evolutionary-based 'false alarms', which are grounded in the brain as well as in social interaction.

Psychobiological models are provided in each of the chapters. These highlight the neurocircuitry that mediates the disorder under discussion, and relate this to its role in mediating relevant cognitive-affective functions, the proximal factors (e.g. neurotransmitters systems) involved in mediating relevant psychopathology, and the distal factors (e.g. evolutionary basis) that may also play a role in accounting for the symptoms. While a broad range of data is presented, the focus is on developing an integrative conceptual approach rather than on highlighting any particular empirical work.

This introductory chapter provides a brief overview of some of the specific advances and impediments in the fields of depression and anxiety, before going on to discuss some of the general conceptual problems that exist for those attempting to build models of psychiatric disorders. The discussion leads in turn to a rationale for the approach, based in cognitive-affective neuroscience and evolutionary theory, that is taken in this volume.

Advances and impediments

Several recent advances in understanding and treating depression and the anxiety disorders seem particularly important (Table 1.1). First, a little over two decades ago, psychiatric classification took a crucial step forward by separating out the broad category of 'anxiety neurosis' into distinct anxiety disorders, each defined in terms of reliable diagnostic criteria (DSM; American Psychiatric Association, 1980). The current nosology has gained international acceptance, and has been useful in ensuring that the reliability for diagnoses of mood and anxiety disorders is at least as high as that for general medical disorders. This volume focuses on major depressive disorder, as well as on the most important anxiety disorders in clinical practice.

Second, the availability of diagnostic criteria for mood and anxiety disorders in turn allowed for rigorous epidemiological surveys to determine the prevalence, comorbidity, and morbidity of these conditions. Clinicians now recognize that depression and the anxiety disorders are not only amongst the most prevalent of all psychiatric disorders (Kessler et al, 1994; Robins et al, 1984) but that they are amongst the most disabling of all medical disorders. The costs of the anxiety disorders alone have been estimated at over 40 billion dollars per year in the United States (Dupont et al, 1996; Greenberg et al, 1999).

Third, there have been significant advances in understanding the psychobiology of depression and the anxiety disorders. Neurochemistry,

Table 1.1 Advances and limitations of current work on depression/anxiety.

	Advances	*Limitations*
Classification/ diagnosis	Recognition of separate disorders, high reliability of diagnosis	Validity of current diagnostic classification system can be questioned
Epidemiology/ morbidity	Awareness of prevalence, morbidity of depression/ anxiety	Continued underdiagnosis, undertreatment of depression/ anxiety
Psychobiology	Depression/anxiety disorders underpinned by specific psychobiology	Precise etiology of depression/ anxiety disorders remains undetermined
Treatment	Introduction of effective pharmacotherapy/ psychotherapy interventions	Limited effectiveness (real world) data, particularly on comorbidity, maintenance issues
Consumer advocacy	Consumer organizations that help destigmatize mood/anxiety disorders	Relatively low levels of mental health literacy and of mental health funding

neuroanatomy, neuroimmunology, and neurogenetics are all rapidly advancing fields, and findings have quickly been applied to understanding clinical disorders. There have also been fundamental advances in a range of cognitive sciences (Gardner, 1985; Gazzaniga, 2000), including cognitive psychology, that have a bearing on understanding depression and anxiety disorders. Modern technologies such as gene sequencing and functional brain imaging offer particular promise for the future.

Fourth, there have been important advances in the treatment of depression and the anxiety disorders. A considerable number of new medications have received regulatory approval for use in some of these conditions, and there is also a larger literature addressing off-label indications for the use of these agents. Furthermore, a number of manualized psychotherapeutic interventions, and cognitive-behavioral psychotherapies in particular, have been found useful in the mood and anxiety disorders. There is a growing database of controlled trials, and the clinicians have a range of effective interventions at their disposal.

Fifth, consumer advocacy has helped to reframe the psychiatric disorders. Psychiatry has arguably long been guilty of some paternalism when it comes to its view of psychiatric patients. Anti-psychiatric critics (Sedgwick, 1982), however, by downplaying the reality of psychiatric conditions have also ignored the specific interests of people who suffer from psychiatric disorders. Increasingly, people with depression and anxiety disorders have formed organizations which aim to destigmatize these conditions and to encourage early diagnosis, appropriate treatment, and additional research. It is encouraging to see the strong move towards physician–patient collaboration in the fight against mental illness.

Nevertheless, there are also major impediments in the field of depression and anxiety disorders, and these too deserve acknowledgment and discussion (Table 1.1). First, classification systems for diagnosing depression and anxiety disorders have received considerable criticism. Although the DSM system allows reliable diagnosis, there is far less evidence that its diagnostic constructs have validity. Thus, for example, the boundaries between normality and abnormality remain poorly defined (Spitzer and Wakefield, 1999), and the categorical nature of DSM constructs seems at odds with more dimensional clinical phenomena. More importantly, perhaps, there seems to be little overlap between clinical symptoms and biological dimensions (van Praag, 1998).

Second, despite increased recognition of the prevalence and morbidity of depression and anxiety disorders, underdiagnosis and undertreatment remain significant problems. The National Comorbidity Survey demonstrated that only a small minority of people who suffer from mood and anxiety disorders receive appropriate care (Kessler et al, 1994). Unfortunately, patients with these conditions are more likely to be high utilizers of medical care. Patients who present with somatic (Kirmayer et al, 1993) and anxiety (Ormel et al, 1991) symptoms are particularly likely to be misdiagnosed in primary care.

Third, despite advances in pathophysiology, we remain ignorant of the precise etiology of depression and the anxiety disorders. In the next section we note that there is currently little evidence that neurotransmitter dysfunction per se is responsible for causing depression and anxiety disorders. Indeed, although we have evidence about multiple factors that may contribute to these disorders, we understand too little about their ultimate causes.

Fourth, despite advances in efficacy research, there is a relative lack of effectiveness research. Efficacy research is limited by a number of factors: subjects fall within a narrow range of symptom severity, they lack comorbid disorders, and the treatment setting is unrepresentative. There is a relative lack of effectiveness or real-word studies in naturalistic settings. Current regulations allow for the introduction of medications to the market after efficacy is shown in the short term; it is crucial to determine the costs/benefits of long-term use. Even simple questions, like which augmentation strategy to choose first in depression, have received vanishingly little prospective research attention (Fawcett et al, 1999).

Fifth, although consumer advocacy in psychiatry has grown, overall levels of mental health literacy are far from satisfactory (Jorm, 2000). All too often there is stigmatization of psychiatric disorders, with patients being blamed for their illness, and with treatment techniques such as 'pull up your socks', or 'get it off your chest' being advised both by the public and by primary care practitioners. 'Natural' remedies are advised (in favor of factory-made pharmaceuticals), with little understanding that plants contain pharmo-active substances with effects and side-effects. The principle that depression and anxiety disorders have relatively high placebo response rates and that controlled studies are needed before treatments can be recommended with any confidence is not well understood.

This volume hopes to provide an approach that is based not only on advances in the fields of depression and anxiety, but that at the same time addresses some of the lacunae that remain.

Conceptual problems

Although there have been significant advances in relation to depression and anxiety disorders, there are also multiple and important gaps. This arguably reflects a major general difficulty that exists in approaching psychiatric disorders: the lack of a rigorous conceptual framework for understanding the nature of these conditions. When clinicians and researchers articulate their underlying concepts and assumptions, it seems that these can vary significantly (Stein, 1991).

Certainly, different models of psychopathology have dominated at different times during the past century or so. A psychodynamic approach, for example, was particularly influential amongst clinicians for many decades. Freud and his followers made a tremendously important contribution to depression and anxiety disorders by providing detailed clinical descriptions, by insisting that a scientific approach to the study of these conditions was possible, and by explaining symptoms in terms of patients' past history and experience.

Nevertheless, many would argue that a hydraulic model of the brain–mind has become too dated to be useful. Psychodynamic clinicians may argue that relational and interpersonal theories have replaced early hydraulic models, so that in practice an acceptance of early Freudian theory is no longer relevant. Nevertheless, for researchers, it is difficult to undertake empirical work when constructs cannot be operationally defined or are inconsistent with current work in other fields on brain and mind (Stein, 1997).

In departments of academic psychology, however, behavioral and cognitivist approaches have ruled. Early behavioral approaches to depression and anxiety disorders provided fundamental contributions; ideas about fear conditioning remain crucial, and exposure techniques are a cornerstone of modern treatments. Nevertheless, a strict behaviorist approach, which focuses solely on behavioral inputs and outputs, viewing the mind as a 'black box' that is inaccessible to scientific study, now has few supporters.

Indeed, the development of computational models of the mind led to a 'cognitive revolution' which replaced the 'black box' model with a focus on cognitive processing (Gardner, 1985). The new field of 'cognitive science' had immediate appeal in so far as it was a multidisciplinary endeavor, theoretically sophisticated, and empirically rigorous. Early cognitive models were based on the linear computers of the time, and led to a range of influential clinical applications, including the characterization and correction of the cognitive dysfunctions characteristic of depression and anxiety disorders (Beck, 1967).

However, when attempts are made to apply the constructs and methods of cognitive science to clinical disorders, significant problems may arise. The majority of work in cognitive science focuses on cognition rather than affect, and on simple laboratory tasks rather than the complex kinds of phenomena seen in the clinic. Perhaps more important, however, is the cognitive science assumption that mental functions can be specified independently of substrate. The view that silicon and brain are logically equivalent turns out not to be particularly useful in building models of mental dysfunction (Stein and Young, 1992).

In psychiatry, an increasingly influential paradigm focuses simply on neurobiology and pharmacology, and, more recently, on molecular neurobiology. Advances in human genomics and proteomics have further strengthened the appeal of this approach. The discovery of the serotonergic system, the delineation of different serotonin receptor subtypes, and the introduction of a range of serotonergic medications, for example, constitute a tremendous advance; there is a temptation for clinicians to reduce psychiatric disorders to 'serotonergic dysfunction' and for researchers to focus on related 'biological markers'.

Such an approach runs a risk of being overly reductionistic. It appears increasingly unlikely that complex psychiatric disorders will ultimately be explicable in terms of any single neurobiological factor or system. Even if all the interactions between different systems and between different components of any one system are fully delineated, there remains the crucial problem of relating such data to behavioral and psychological phenomena. Psychiatric conditions, including the depression and anxiety disorders, are emergent phenomena and will require correspondingly complex explanatory accounts.

All of these different models – the psychodynamic, cognitive-behavioral, and biological – have important similarities: they can perhaps be grouped under the rubric of a 'clinical' approach (Table 1.2). Such an approach has a long conceptual history. It depends on a view of science that highlights the importance of collating objective data about the world-out-there, and of finding laws which cover the relationships between such data (Bhaskar, 1978). It is also consistent with a classical view of psychology which states that mental phenomena can similarly be operationally defined, and explained by the kinds of scientific laws that operate in the physical sciences. And finally, the clinical approach is consistent with a view of medicine that emphasizes the importance of 'objective' diseases and of using scientific methods to explain them.

Indeed, given the difficulties with each of the approaches outlined above, one possibility is to conclude that the clinical approach has fundamental flaws. An important paradigm argues that psychiatric disorders and symptoms cannot be understood within the framework of natural and medical science, but rather require an approach that falls within the human and social sciences. Certainly, lay people have an immediate tendency to try and understood mood and anxiety symptoms in terms of their social context: as reactions to loss, trauma, or other stressors. And there is a sophisticated literature that similarly argues that depression and anxiety cannot simply be reduced to medical disorders; this is an erroneous and unwarranted extension of the medical model (Sedgwick, 1982).

This 'hermeneutic approach' (Table 1.2) again has a long conceptual history. It depends on a view of science as being simply one kind of social activity, rather than a privileged approach to the truth (Bhaskar, 1978). It is also consistent with an influential view of psychology which highlights the role of interpretation and understanding in making sense of mental states, and which draws a sharp distinction between the methods of the psychological and physical sciences. Similarly, situated cognitivists argue that mental phenomena cannot be specified independently of their social context (Norman, 1993). And finally, the hermeneutic approach is consistent with a view of medicine that highlights the importance of 'illness': the subjective expression and experience of suffering from a disorder (Kleinman, 1988).

Such an approach is unlikely to appeal to researchers interested in

Table 1.2 Conceptual approaches to depression/anxiety.

Approach	*Science*	*Psychology*	*Medicine*	*Depression/anxiety*
Clinical	Discover laws that cover objective phenomena	Discover laws that cover mental phenomena	Discover laws that cover pathological phenomena	Depression/anxiety can be defined in terms of DSM categories
Hermeneutic	Science is not simply objective, but is also a social practice	Psychology involves interpretation and understanding	Medicine must include illness – the expression and experience of disorder	'Depression/anxiety' represent meaningful responses to social context
Embodied	Science is a social practice that uncovers real mechanisms	Psychology involves not only explanation but also understanding	Medicine needs to recognize both disease and illness	Depression/anxiety must be understood in terms of proximal/distal mechanisms, embedded in society/brain

using natural science methods to explain psychiatric disorders. Nevertheless, a number of clinicians have felt sufficiently attracted to the hermeneutic perspective to argue that, at the least, all patients should be understood from several different perspectives (McHugh and Slavney, 1988). Similarly, a 'biopsychosocial' model receives a great deal of lip service. Nevertheless, the question of how to integrate different perspectives also requires a detailed answer. In the next section, one possible approach, drawing on the modern field of cognitive-affective neuroscience, is suggested.

Cognitive-affective neuroscience

The various clinical and hermeneutic approaches might seem like so many theoretical 'straw men'; but for the fact that they retain considerable influence in practice. Nevertheless, it does seem that a more integrated approach is gradually being formulated and accepted. Cognitive neuroscience or (perhaps more appropriately in the case of psychiatry) cognitive-affective neuroscience, for example, has its roots in cognitive science but, as with many more recent computational models (e.g. neural networks), is particularly focused on the interface with neuroscience.

Cognitive-affective neuroscience may have the potential to incorporate many of the advances of cognitive science (Stein and Young, 1992), but also to integrate these with developments in neuroscience as well as evolutionary science. The neural structures that react to appetitive and aversive stimuli, and that form the foundation of human emotion, have evolved over millennia. A non-reductionistic cognitive-affective neuroscience may also be able to draw on the strengths of both the clinical and hermeneutic approaches, and so allow a synthetic or 'embodied' approach to psychopathology.

Thus, the different chapters of this volume attempt to provide cognitive-affective neuroscience models of depression and each of the anxiety disorders. The term 'cognitive-affective neuroscience' underscores the attempt to address the brain–mind using an integrated scientific approach, an attempt that is particularly relevant for models of psychopathology. One of the themes of this volume, for example, is that depression and anxiety disorders can be conceptualized in terms of evolutionary-based 'false alarms', which in turn must be understood in terms of their social and brain context.

Before proceeding to consider these different disorders, there are a few conceptual aspects of the approach here that can be highlighted in order to contrast it with the clinical and hermeneutic positions described earlier (Table 1.2). First, a synthetic approach defines science in terms of an effort to understand the underlying mechanisms that account for the phenomena under study. In the case of psychiatric disorders, such mechanisms are likely to involve both brain factors and sociocultural factors, and they are both proximal (for example, the particular subreceptors of the serotonin (5-HT) system may play a part) and distal (for example, the evolutionary role of 5-HT function can be relevant).

Such a view follows the hermeneutic approach in allowing that science is a social practice, but it also argues that science can uncover the real mechanisms that generate the phenomena of the world (Bhaskar, 1978). Similarly, a synthetic approach to psychology holds that this discipline demands a focus on both explanation and understanding, and sees mind as an emergent property of the brain. Cognitive psychology must address the embodiment of mental states in both social contexts and in the brain. Finally, this perspective argues that medicine must engage with both disease (biomedical disorders) and illness (the expression and experience of such conditions). A cognitive-affective neuroscience perspective on depression and the major anxiety disorders is outlined in each of the next chapters.

Each chapter begins by considering the neurocircuitry that mediates normal cognitive-affective processes. It then discusses the neurotransmitters that may play a particularly important role in the dysfunction of such circuits. Given the important role of serotonin in recent research on mood and anxiety disorders, this system is highlighted. Indeed, before going on to each of the individual chapters, it may be useful to consider briefly the serotonin system as a useful 'thing-with-which-to-think' (Pappert, 1980), in order to compare and contrast the clinical, hermeneutic and embodied approaches (Table 1.3).

Table 1.3 Conceptual approaches to the serotonin system.

Clinical approach	There is an invariant relationship between the presence of serotonin dysfunction and that of psychiatric disorder. The serotonin system can be conceptualized as 'hardware' (with a focus on nature) or 'software' (with a focus on nurture).
Hermeneutic approach	Description of serotonergic changes is irrelevant to understanding the expression and experience of psychiatric disorder. Computational models of psychopathology cannot be taken literally.
Integrative approach	Normal mood/anxiety regulation, as well as 'false alarms' of this system, are mediated by the serotonin system. Serotonergic function is embedded in social contexts and in the operation of neurons; a unique computational metaphor ('wetware') is needed to describe it.

A clinical view of serotonin takes the view that depression and anxiety disorders are 'serotonergic disorders'. Data on serotonergic markers or on response of these conditions to serotonergic agents are taken to point to the causal role that the serotonin system plays in their pathogenesis. There are several problems with this view: it oversimplifies the etiology and treatment of depression and anxiety disorders, and appears to reduce (erroneously) complex cognitive-affective states to a simple neurochemical formula. Although such a position is arguably a 'straw man' in the sense that no scientist would take this view, it may be an influential one (a lay view of depression as a serotonin imbalance, for example, is common).

A hermeneutic view, however, may argue that by focusing on serotonin in depression and anxiety disorders we are missing the target by a considerable distance. Certainly, a number of 'anti-psychiatry' authors have argued that the use of selective serotonin reuptake inhibitors (SSRIs) constitutes a disservice to psychiatric patients. A more sophisticated point is that despite the widespread use of these agents, we have very little understanding of how they work, and that there is also relatively little evidence that they work by regulating the serotonergic system (Hyman and Nestler, 1996). Nevertheless, a hermeneutic view seems to ignore an important level of biological reality. Post-modern critiques of psychiatry have themselves become a literary genre that ignores important aspects of the objective world.

A synthetic perspective on serotonin emphasizes that normal mood and anxiety are regulated by a range of neurotransmitter and neurochemical systems including that of serotonin. The proximate mechanisms (that is the psychobiological factors) underlying abnormal mood and anxiety remain poorly understood, but appear to be mediated in part by this complex monoamine system. Serotonergic neurons branch throughout the brain and are therefore able to exert a range of different effects (Figure 1.1). The ultimate mechanisms (that is, the evolutionary

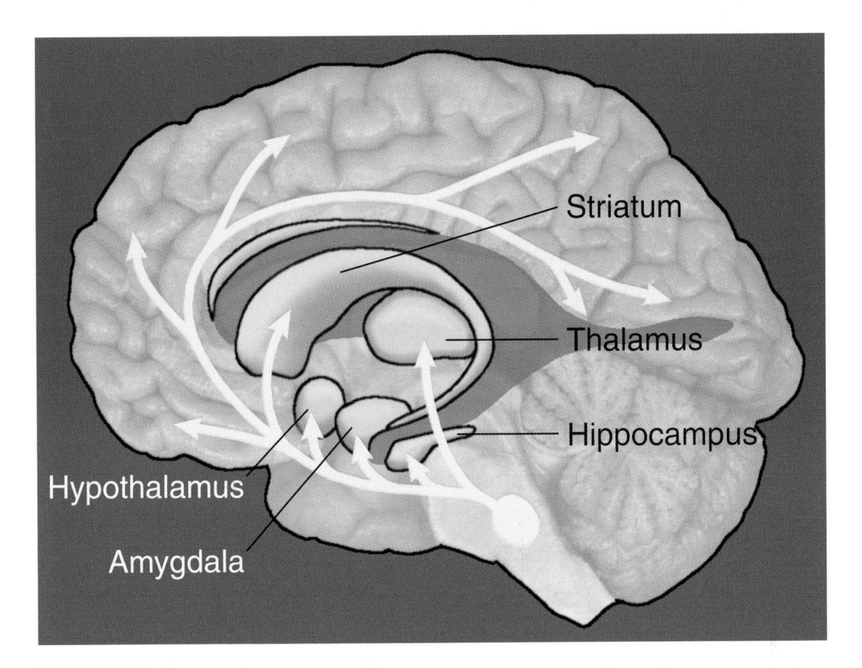

Figure 1.1 *Serotonergic neurons branch widely: neurons originate in the raphe nucleus and branch extensively to the amygdala, hippocampus, hypothalamus, striatum, cingulate, prefrontal and frontal cortex.*

derivation) of the serotonin system and of mood and anxiety disorders are perhaps even less well understood, although several new insights have been made in recent years.

The serotonin system is neither 'hardware' nor 'software' – rather it is unique in being a form of 'wetware' that is embedded within complex social contexts (for example, changes in social hierarchies influence serotonin activity (Raleigh et al, 1984, 1985)), as well as in the brain (mood and anxiety are states that cannot be fully captured in a symbolic computational model). Just as changes in social hierarchy lead to brain changes, so specific psychotherapeutic interventions can alter functional neuroanatomy (see Chapter 4). Also, a cognitive science attempt to simulate computationally such phenomena as depression/anxiety must address the way in which these states are grounded (Harnad, 1991) in neuroanatomical circuits and their neurotransmitters (such as serotonin).

In highlighting this kind of conceptual issue in a clinical book there is the risk of being criticized by philosophers (for being overly clinical) and by clinicians (for being overly philosophical). However, by addressing this kind of issue, certain points of discussion for conceptual science (philosophy) that might otherwise be ignored are raised, and simultaneously an effort is made to bring some conceptual rigor to areas which may be oversimplified in clinical writing. As clinicians, we sometimes forget that our theories are based on philosophical approaches to science in general and the mind in particular.

Thus, in this volume an attempt is made to incorporate the best of both the clinical and the hermeneutic approaches. We argue that psychiatric science has reached the point where a cognitive-affective neuroscience for the clinic is possible. This framework allows both explanation and understanding, and is able to address the embodiment of mind in both society and in brain. Inclusion of evolutionary theory allows both proximate and distal mechanisms of psychopathology to be addressed. Put differently, this volume begins with the idea that an integrated approach to the brain–mind is not only possible but also desirable. In the remaining chapters, we attempt to describe depression and the anxiety disorders using this kind of integrated approach.

chapter 2

Major depression

Symptoms and assessment

Major depression (major depressive disorder) is one of the most prevalent of all psychiatric disorders (Kessler et al, 1994), and one of the most disabling of all medical disorders (Murray and Lopez, 1996). It is a disorder seen in all age groups, both genders (although it is more common in women), and throughout the world. The complication of suicide remains a perennial worry for clinicians, and an important rationale for rigorous treatment of depression.

The symptoms of depression are both cognitive and somatic (Table 2.1). Arguably at the core of depression is the phenomenon of anhedonia – a decrease in the ability to experience enjoyment and pleasure. Cognitive symptoms include negative thoughts about the self, world, and future (Beck, 1967). Somatic symptoms include increased or decreased appetite (changes in feeding), increased or decreased energy and interest in the environment (changes in foraging), irritability and hostility (changes in fighting), and decreased libido. Notable symptoms include increase or decrease in psychomotor activation, and changes in concentration and memory.

Assessment of an episode of major depression requires evaluation of the severity and subtype of depression, of comorbid disorders, of suicidal ideation and disability, and of psychosocial stressors and supports. The differential diagnosis of major depressive disorder from bipolar depression

Table 2.1 Symptoms of major depression (modified from DSM-IV-TR).

(A) Five (or more) of the following symptoms have been present during the same 2-week period and represent a change from previous functioning; at least one of the symptoms is either (1) depressed mood or (2) loss of interest or pleasure:

- depressed mood most of the day, nearly every day
- markedly diminished interest or pleasure in all, or almost all, activities most of the day, nearly every day
- significant weight loss when not dieting or weight gain (e.g. a change of more than 5% of body weight in a month), or decrease or increase in appetite nearly every day.
- insomnia or hypersomnia nearly every day
- psychomotor agitation or retardation nearly every day
- fatigue or loss of energy nearly every day
- feelings of worthlessness or excessive or inappropriate guilt (which may be delusional) nearly every day
- diminished ability to think or concentrate, or indecisiveness, nearly every day
- recurrent thoughts of death (not just fear of dying), recurrent suicidal ideation without a specific plan, or a suicide attempt or a specific plan for committing suicide

(B) The symptoms cause clinically significant distress or impairment in social, occupational, or other important areas of functioning.

(C) The symptoms are not due to the direct physiological effects of a substance (e.g. a drug of abuse, a medication) or a general medical condition (e.g. hypothyroidism).

(D) The symptoms are not better accounted for by bereavement.

is crucial; in this volume, we focus only the former condition. In addition, general medical disorders and substance use must be excluded as potential causes of depressive symptoms.

Useful scales in measuring depression symptoms include the Montgomery–Åsberg Depression Rating Scale (Montgomery and Åsberg, 1979)

(Table A.2, see Appendix). This 10-item scale has become increasingly popular, not only because it allows a reliable measure of symptom severity, but because it is easily and quickly administered and demonstrates good sensitivity to change during treatment.

Cognitive-affective considerations: regulation of mood/processing of reward

A broad range of cognitive-affective functions may be relevant to understanding depression, and may correspondingly suggest an initial approach to the neurocircuitry of this disorder. Executive functions, including modulation of emotional behavior, for example, would seem relevant to understanding cognitive features (e.g. depressive schemas) that characterize depression, and are mediated by frontal cortex. The construct of low positive affect, which may be relevant to depression, has been associated with hypoactivation of left frontal cortex (Mineka et al, 1998).

Similarly, emotional memory systems are relevant to depression, suggesting amygdala and hippocampal involvement. Depressed patients show a bias towards recalling negative information, particularly when it is self-referential. This occurs during both explicit memory tasks and implicit tasks (when memory is tested indirectly) (MacLeod and Byrne, 1993). Psychomotor function may be relevant to understanding motoric symptoms of depression, and is mediated by striatal circuits. Feeding, foraging, and so forth, are related to the somatic symptoms of depression, and suggest involvement of the hypothalamus and hypothalamus–pituitary–adrenal (HPA) axis. These various circuits are themselves interlinked.

Much of our understanding of these cognitive-affective phenomena and their neurocircuitry is based on animal studies. Of course, there are limitations to such work, particularly in addressing more complex and abstract phenomena that are unique to humans. Nevertheless, there is a broad preclinical literature showing that pathways linking limbic structures (amygdala, hippocampus, hypothalamus) to a network of frontal, paralimbic (ventral frontal, cingulate, insula, anterior temporal poles), striatal, and brainstem regions are crucial in mediating affective and motivational behavior (Damasio, 1996; MacLean, 1949; Rolls, 1990). By extrapolation, similar regions are likely to be involved in human mood regulation and reward mechanisms.

Functional imaging studies using normal humans are potentially useful in shedding additional light on cognitive-affective phenomena relevant to depression, and their mediating neurocircuitry. Several studies of transiently induced sadness in normal subjects have confirmed the role of a broad network of cortical-limbic circuits in the modulation of mood, and in evaluating the emotional significance of stimuli (Lane et al, 1997; Mayberg et al, 1999). Ultimately, further work is needed to delineate the contribution of particular neuronal circuits to normal regulation versus psychopathology, and also to state versus trait characteristics (Davidson, 1994; Drevets, 2000).

Neurocircuitry of major depression

In approaching the neurocircuitry responsible for mediating any psychiatric disorder, an important initial approach, highlighted throughout this volume, is the careful study of patients with symptoms secondary to general medical disorders. Seminal work by Robinson and colleagues, for example, demonstrated that in patients with stroke, left-sided lesions were more likely to be associated with depression, while right-sided lesions were more likely to be associated with mania. Notably, in depressed patients with left-sided lesions, lesions closer to the frontal pole were associated with more severe depression (Robinson et al, 1984). Conversely, patients with secondary depression have evidence of frontal dysfunction (Mayberg, 1994), as do patients with late-onset depression (in which neurological insults may be relatively important) (MacFall et al, 2001).

A range of other findings from the literature on depression secondary to general medical disorders is relevant to developing a neuroanatomical model of this disorder (Byrum et al, 1999). In support of the role of striatal neurocircuitry in depression, classical observations include the association between lesions in these circuits (e.g. Parkinson's disease, vascular depression) and depressed mood, and between depression and psychomotor disturbance (Sobin and Sackheim, 1997). Furthermore, it is notable that patients with various abnormalities of the hypothalamus and HPA axis may suffer profound depression.

Another useful approach to the psychobiology of depression has been to focus on the sequelae of adverse events, particularly of early adversity. Such work has its roots in the seminal observations that primate separation ultimately results in a 'depressive' picture (Bowlby, 1980). Recent work has documented that the sequelae of early adversity, in both animals (Sanchez et al, 2001) and humans (Heim and Nemeroff, 2001), include altered neuroendocrine function and (perhaps as a consequence of this) decreased hippocampal volume.

A range of other studies has further contributed to delineating the neurocircuitry of depression. Postmortem morphometric studies, for example, have found evidence of atrophy in specific cell regions (Duman et al, 2000). Demonstration of specific cognitive deficits in depression has provided clues about the neurocircuitry of depression (Austin et al, 2001). Neurosurgery is occasionally used in the treatment of refractory depression, again shedding light on underlying neurocircuitry. Brain imaging has, however, proved key for advancing this area.

Structural imaging studies, for example, have identified abnormalities in prefrontal, limbic/paralimbic (cingulate, hippocampus), and striatal regions in depressed patients (Sheline, 2000). Functional imaging studies similarly suggest that depression is characterized by decreased activity across this range of interconnecting neurocircuits (Videbach, 2000) (Figure 2.1). However, these conclusions remain somewhat tentative; a number of studies of depression have, for example, found not only hypoactivity (e.g. in neocortical areas) but also areas of hyperactivity (e.g. in paralimbic regions (Mayberg et al, 1999) or the amygdala (Drevets, 2000)).

There are relatively few imaging studies of psychotic mood disorders. A recent review suggested that when overlapping regions are involved, the dysfunctions in psychotic mood disorders are more severe than those in nonpsychotic depression (Wang and Ketter, 2000). Interestingly, a number of studies suggest that in mania there is heightened

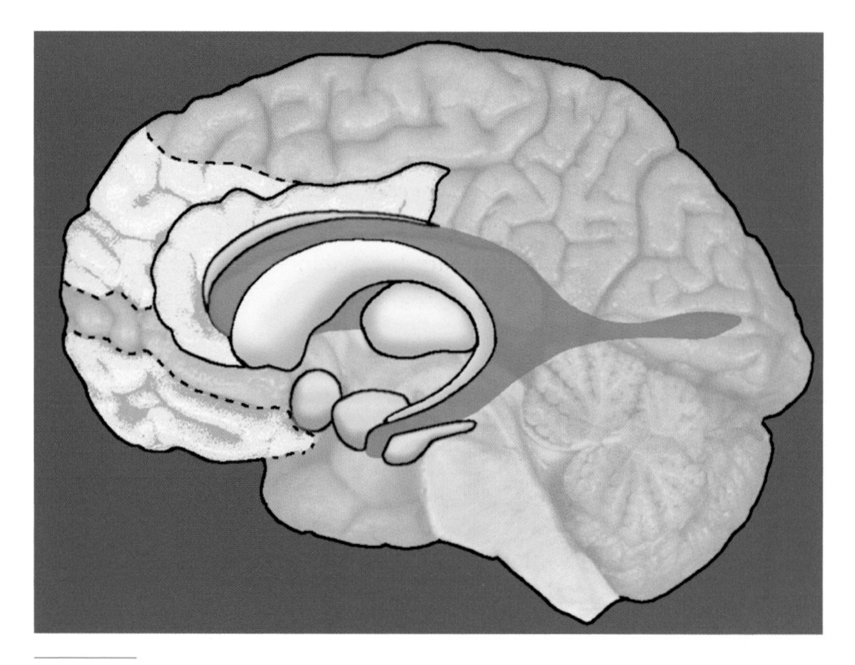

Figure 2.1 *Functional neuroanatomy of depression: decreased activity in prefrontal, paralimbic (cingulate), and striatal regions.*

activity in a cortical–subcortical neural circuit that includes the anterior cingulate and caudate (Blumberg et al, 2000).

Clearly, further work is needed to consolidate a neuroanatomical model of depression. Differences in findings across functional imaging studies may ultimately be explicable in terms of methodological variation or in terms of subject heterogeneity. Additional research is needed to delineate the neuroanatomy of different subtypes of depression, and to correlate functional neuroanatomy with clinical dimensions (Bench et al, 1993). More work is also needed to replicate early findings that functional findings predict pharmacotherapy response (Mayberg et al, 1997).

A number of studies have reported that SSRIs and other agents act to normalize functional imaging in depressed patients, although exact findings are inconsistent (Kennedy et al, 2001). Exciting research also suggests that there is an overlap in the effects of pharmacotherapy and psychotherapeutic intervention (Brody et al, 2001; Martin et al, 2001), with placebo response mediated by different paths (Leuchter et al, 2002). Cognitive therapy may be mediated by a 'top-down' cortical influence on limbic pathways, whereas neurosurgical lesions (limbic leucotomy or subcaudate tractotomy) (Malhi and Bartlett, 2000) comprise a 'bottom-up' attack on the limbic system. In this view, pharmacotherapy can be understood as 'mixed', with primary brainstem–limbic effects, but also secondary cortical effects (Mayberg et al, 1999).

Neurocircuitry and neurochemistry

A broad range of studies has demonstrated abnormalities of peripheral serotonergic markers in depression, with the most widely reported abnormality involving decreased serotonin transporter (5-HTTP) binding (Owens and Nemeroff, 1994). In addition, a consistent finding in both preclinical and clinical studies has been of an association between decreased serotonergic function and impulsivity (including aggression and suicide (Stein et al, 1993)). At postmortem, depression may be characterized by reduced numbers of dorsal raphe neurons (Baumann et al, 2002). Furthermore, functional imaging and postmortem studies have confirmed decreased serotonin transporter (5-HTTP) binding in depression/suicide (Malison et al, 1998; Mann et al, 2000). 5-HTTP binding, for example, was decreased in prefrontal cortex of subjects with a history of depression, with binding lower in ventral prefrontal cortex in suicides (Mann et al, 2000).

Dynamic studies of the 5-HT system in response to serotonergic agonists, or to serotonergic depletion, offer methodological advantages over studies of static measures; a range of data again points towards aberrant serotonergic neurotransmission in depression (Charney, 1998). Depletion studies, for example, have shown that decreased 5-HT synthesis precipitates symptoms of depression in healthy and in remitted depressed patients (Bell and Nutt, 2001). In relapsed patients (but not in unaffected subjects), positron emission tomography (PET) scanning shows a decrease in activity in dorsolateral prefrontal cortex, orbitofrontal cortex, and thalamus (Bremner et al, 1997b). The link between hyposerotonin function, hypofrontality, and mood dysfunction is again apparent.

A currently influential hypothesis is that SSRIs exert their effects by desensitization of 5-HT_{1A} somatodendritic autoreceptors (which initially serve as a 'brake' on serotonin neurotransmission). This accounts for the relatively slow time course of pharmacotherapy response in depression, with only gradual desensitization and increased neurotransmission (as the 'brake' is effectively released) (Stahl, 1998). Ultimately there may be increased serotonergic activity, and reversal of dysfunction (Figure 2.2). There is a growing literature on the molecular imaging of 5-HT receptor subtypes at baseline, and after SSRI treatment, with some evidence, albeit inconsistent, in favor of such a hypothesis (Becker et al, 2001; Sargent et al, 2000; Staley et al, 1998).

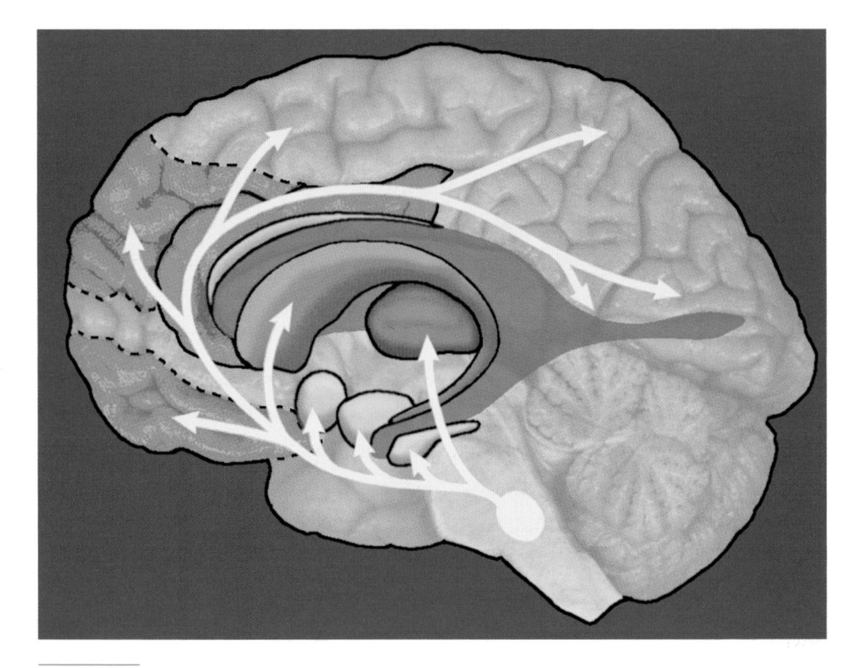

Figure 2.2 *Effects of SSRIs on functional neuroanatomy of depression: normalization of activity in prefrontal, paralimbic (cingulate), and striatal regions.*

There is also growing interest in the relationships between serotonin genotype, early experience, functional imaging, and pharmacotherapy response. Primate research, for example, suggests that 5-HTTP genotype and early rearing environment interact to account for behavior (Lesch, 2001). In the clinical setting it might be hypothesized that the low-activity 5-HTTP genotype is associated with greater impulsivity, hypofrontality, and worse response to treatment with SSRIs. To date findings are inconsistent (Mann et al, 2000); any single gene may only account for a relatively small amount of the variance when considering this kind of complex association. However, the mapping of the human genome, together with the increasing availability of molecular imaging (Staley et al, 1998), lay the ways for future advances in this area.

Several additional neurochemical systems also play an important role in mediating depression. The noradrenergic system, for example, may play a particularly important role in mediating abnormal drive and energy in depression. Attempts to delineate more 'serotonergic' and more 'noradrenergic' symptoms of depression have received renewed attention since the introduction of specific noradrenaline reuptake inhibitors.

The dopamine system is thought to play a crucial role in processing motivation and reward, and may also play a role in depression. Interestingly, a range of antidepressants potentiate dopamine transmission in the nucleus accumbens (Serra et al, 1992). Dopamine hypofunction may be particularly important in patients with psychomotor retardation; decreased striatal dopamine uptake is present in retarded depressed patients compared with non-retarded patients and healthy controls (Martinot et al, 2001).

Certainly such work often serves as a useful theoretical foundation for clinical psychopharmacology. Thus, in the management of depression, SSRIs can be (i) accelerated (for example, by augmentation with 5-HT_{1A} partial agonists that act to further desensitize somatodendritic autoreceptors); (ii) augmented (for example, by adding a moiety or a drug with post-synaptic serotonergic effects); or (iii) complemented (for example, by adding moieties or drugs with dopaminergic or noradrenergic effects).

Important additional neurochemicals that can be targeted by novel antidepressants include neuropeptides (such as corticotrophin release factor, substance P), excitatory amino acids (e.g. glutamate), and gases (e.g. nitric oxide). The neurogenetics and neuroimmunology of depression are also rapidly evolving research fields. In future years, clinicians will hopefully have access to a range of new pharmaceutical agents that involve first messenger systems other than the traditional monoamine neurotransmitters (Lesch, 2001).

Further, biological factors distal to the cell surface are crucial. There is a growing appreciation of the role of second messenger systems (G-proteins, kinases), third messengers (e.g. transcription factors such as CREB), and ultimately target genes (e.g. the serotonin transporter gene, neurotrophins like brain-derived neurotrophic factor) in mediating depression. Ultimately, the response of antidepressants must be understood at a genetic level (Hyman and Nestler, 1996).

Evolutionary considerations

Evolutionary medicine attempts to supplement standard accounts of the proximal mechanisms involved in disease, with hypotheses about the evolutionary origins of pathology (Nesse and Williams, 1994). Although aspects of this framework have long been in existence (Darwin, 1965), in recent years there has been renewed theoretical interest as well as growing empirical work. The work of Nesse (Nesse, 2000) is perhaps the most sophisticated evolutionary-based account of depression published to date.

In his view, depression is evolutionarily advantageous in the face of a situation where continued effort to pursue a major goal will likely result in danger or loss. Such situations include, for example, a fight with a dominant figure, or the failure of a major life enterprise. In these cases, pessimism and lack of motivation are useful in so far as they inhibit actions which may be dangerous (e.g. attempting to do battle with a much stronger figure) or wasteful (e.g. attempting to start a new enterprise without adequate resources).

Such an approach may be useful in providing a distal account on which to base other kinds of psychobiological knowledge (Duchaine et al, 2001). In this context it is interesting to note that preclinical and human studies indicate that changes in serotonergic transmission mediate thresholds for adopting passive or waiting attitudes, or accepting situations that necessitate or create strong inhibitory tendencies (Soubrie, 1986). More generally, it has been argued that serotonin facilitates gross motor output and inhibits sensory information processing (Jacobs and Fornal, 1995).

In the future we can expect that 'Darwinian psychiatry' (McGuire and Troisi, 1998; Stevens and Price, 1996) becomes supported by data, and is in turn increasingly used as a rationale for particular kinds of intervention. Nevertheless, in the interim, it is important not only to acknowledge the speculative nature of such work, but also to emphasize that many subtypes of depression may not be explicable in terms of a meaningful response to environmental stressors.

Certainly, as indicated earlier, depressive symptoms may emerge as the direct consequence of a general medical condition; depression in the context of stroke, for example, may be more powerfully explained in terms of a brain lesion than in terms of loss. More commonly, individual psychobiological variations may contribute an important risk factor for depression, which in turn may often need to be conceptualized as a 'dysfunction' rather than as a functional response. Similarly, evolutionary models of acute depressive episodes may not apply to chronic symptoms; Kraepelin was the first to argue that psychosocial stressors play a greater role in the initial than in subsequent episodes of depressive disorders, and a 'kindling' hypothesis of recurrent depression has received support from epidemiological (Kendler et al, 2000), biological (Post, 1992), and cognitive (Segal, 1996) perspectives. Different subtypes of depression (bipolar depression, seasonal affective disorder, etc.) may reflect different kinds of underlying dysfunction.

Management

An important part of the management of depression involves establishing rapport with the patient, and providing a sense of hope. A cognitive-affective neuroscience perspective reminds us that positive social relationships may not only impact positively on cognitive 'distortions', but that they may even have direct and salutary neurobiological effects. Findings, for example, that touch can modulate neurotransmitter and neuroendocrine effects (O'Donnell et al, 1994) are consistent with a long ethological and evolutionary literature in the importance of nurturant social interaction, and increase our awareness of how important the 'holding environment' of the therapeutic relationship is.

The availability of a broad range of antidepressant agents has significantly contributed to improving the prognosis of depression. There is broad agreement that newer generation antidepressants are better tolerated and easier to prescribe, and these are therefore increasingly seen as comprising the first line of pharmacotherapy (Fawcett et al, 1999). There is also ongoing discussion of whether certain agents are more likely to lead to remission or have a more rapid onset (Gelenberg and Chesen, 2000). While overoptimism about the future treatment of depression should certainly be avoided (van Praag, 1998), a better understanding of the pathogenesis of this disorder may well lead to the development of novel improved agents in the future.

At the same time, a range of different psychotherapies have proven valuable in the treatment of depression. Schemas are cognitive-affective structures which govern the way in which information is assimilated and which, in turn, accommodate the interpretation of this information (Piaget, 1952). Activation of depressive schemas may be associated with an implicit focus on negative memories and altered levels of activity. While pharmacotherapy may be able to alter the prefrontal, striatal, and amygdala-hippocampal circuits which mediate such schemas, psychological and behavioural interventions are also effective. A cognitive-affective neuroscience perspective readily supports the use of pharmacotherapy, psychotherapy, and their combination.

In patients that do not respond to first-line pharmacotherapy and/or psychotherapy, particular attention should be paid to ensuring appropriate diagnosis (excluding bipolarity, and underlying general medical disorders), as well as to the possibility of early trauma and ensuing maladaptive schemas that reduce the efficacy of standard psychotherapies (Stein and Young, 1992). There is a growing awareness that the neurocircuitry mediating depression involves multiple neurotransmitters, and that the pharmacotherapy of treatment-refractory depression may well require the use of medication augmentation and combination. Nevertheless, it is important that appropriate doses and durations of monotherapy be attempted, and that cognitive-behavioural interventions be considered.

Part of the management of depression may well involve liaison with consumer advocacy groups and others involved in broader struggles against the stigmatization of mental disorders and for the early recognition and treatment of these conditions (Jorm et al, 1997). Broad segments of society continue to perceive depression as a spiritual weakness, moral failing, or character flaw. The message that depression is a medical disorder, which like other such conditions is mediated by a complex range of factors, and that requires timely, rigorous and evidence-based intervention, must continue to be spread. Clinicians arguably underestimate the extent to which they can contribute to such advocacy work, and the extent to which such work may be useful for their patients.

Conclusion

Depression cannot simply be reduced to any cognitive formula (somatic factors are too important, a computer cannot be depressed) or to any neurotransmitter dysfunction (sociocultural factors are too important). At the same time, it is unlikely that depression can always be understood simply in terms of a meaningful response to an adverse environment (in the clinical context, depression can often be characterized as an irrational response, a 'false' response).

Indeed, depression can arguably be understood in terms of a model that grounds the symptoms of this disorder in the wetware of the brain–mind and in human social interaction. A concept of brain–mind dysfunction provides an explanatory basis for dissecting the particular cognitive distortions and somatic symptoms that characterize depression. In addition, however, an understanding of social losses, stressors, and supports is crucial for appreciating the experience and expression of depression. Only an integrative approach, providing a framework for a range of data, will have sufficient power to tackle the complexity of this disorder.

In adopting a multidimensional and comprehensive approach to depression, it may be important to focus not only on aspects of psychopathology, but also on resilience. Work on hippocampal shrinkage and neurotransmitter dysfunction in depression should not be taken to imply that depression is an irreversible and refractory illness. Such a conclusion belies the enormous resilience of humans and the powerful effects of modern treatments. Certainly, in both basic and clinical arenas, there is a growing appreciation of neuroplasticity. Future work on depression may well focus not only on how stressors negatively affect neurocircuitry, but also on the neurotrophic effects of antidepressant intervention (Lesch, 2001).

chapter 3

Generalized anxiety disorder

Symptoms and assessment

Generalized anxiety disorder (GAD) is a moderately common anxiety disorder in epidemiological studies (Kessler, 2001) and the most common anxiety disorder in primary care settings (Maier et al, 2000). A broader construct, similar to the older 'anxiety neurosis', is likely to be even more prevalent in the community and in the clinic. GAD is a chronic and disabling disorder, more common in women, with risk for age of onset beginning in the teens and cumulative lifetime prevalence increasingly in roughly linear fashion until the mid forties.

GAD has been a somewhat controversial diagnosis. DSM-III initially characterized GAD as a residual disorder, and subsequent authors have similarly argued that in view of its apparently high comorbidity with other disorders GAD should be conceptualized as a prodrome or severity marker of another condition. Nevertheless, there is evidence that comorbidity is no higher in GAD than in other disorders such as depression, and a growing consensus views GAD as an independent condition characterized by specific psychobiological features (Kessler, 2001).

GAD is characterized by worries that are difficult to control, and by a range of somatic symptoms (Table 3.1). The increased focus in DSM on 'worry' at the expense of 'tension' has been criticized by a number of authors (Rickels and Rynn, 2001). Indeed, in some ways the term 'tension disorder' seems more appropriate: GAD is characterized by both

Table 3.1 Symptoms of GAD (modified from DSM-IV-TR).

(A) Excessive anxiety and worry (apprehensive expectation), occurring more days than not for at least six months, about a number of events or activities (such as work or school performance).

(B) The person finds it difficult to control the worry.

(C) The anxiety and worry are associated with three (or more) of the following six symptoms (with at least some symptoms present for more days than not for the past six months).

1. restlessness of feeling keyed up or on edge
2. being easily fatigued
3. difficulty concentrating or mind going blank
4. irritability
5. muscle tension
6. sleep disturbance (difficulty falling or staying asleep, or restless unsatisfying sleep)

(D) The focus of the anxiety and worry is not confined to features of an Axis I disorder.

(E) The anxiety, worry, or physical symptoms cause clinically significant distress or impairment in social, occupational, or other important areas of functioning.

(F) The disturbance is not due to the direct physiological effects of a substance (e.g. a drug of abuse, a medication) or a general medical condition (e.g. hyperthyroidism) and does not occur exclusively during a mood disorder, a psychotic disorder, of a pervasive developmental disorder.

psychic tension (irritability, insomnia), and by somatic tension (muscle tension, pains, and aches). Such tension may be followed by worry, which can be understood as an avoidance behavior.

Assessment of GAD requires evaluating the course of psychic and somatic symptoms, comorbid mood and anxiety disorders, and impairment in function. Psychosocial stressors and supports must be determined. Although it is clear that GAD is associated with increased medical utilization, it should also be remembered that general medical conditions and substance use may contribute to anxiety symptoms, and these should therefore be excluded.

The severity of GAD symptoms can be rated using the Hamilton Anxiety Rating Scale. This scale was developed as a companion scale to the Hamilton Depression Rating Scale, addressing symptoms thought to be particularly reflective of general anxiety. In clinical practice, it may also be useful to use a scale generated from the DSM-IV criteria for GAD (Table A.2, see Appendix), although psychometric data for this instrument are not yet available.

Cognitive-affective considerations: planning the future

Whereas depression symptoms characteristically focus on the past, GAD symptoms revolve around anticipation of future harm. Indeed, there is an empirical literature documenting that anxious patients show greater anticipation of future negative events (MacLeod and Byrne, 1993). A number of brain regions may conceivably be involved in such anticipation, including prefrontal cortex, given its importance in mediating executive functions such as planning and predicting.

In addition, GAD symptoms presumably involve general activation of what might be termed the 'basic fear circuit'. Anxious patients attend selectively, and not necessarily with awareness, to threatening cues (Macleod and Byrne, 1993). While fear conditioning will be discussed in detail in Chapter 5, for the moment it can be noted that the amygdala plays a crucial role in this fundamental process. It has also been hypothesized that the bed nucleus of the stria terminalis (part of the extended amygdala) is particularly important in free-floating anxiety (Davis and Whalen, 2001). The hippocampus may be particularly relevant in more complex situations involving conflict (Gray and McNaughton, 1996) or avoidance.

There is a growing brain imaging literature focusing on the induction of negative emotions in normal controls. Findings may depend on a range of factors including provocation strategy, but activation of inferior frontal cortical and anterior temporal areas has been documented with different emotions including anxiety (Kimbrell et al, 1999).

Orbitofrontal regions are particularly likely to be activated by internally generated (rather than external) affective representations (Zald and Kim, 1996). Anterior temporal lobe activation during emotional processing may reflect a role in assessment of context, whereas amygdala activation may reflect the experience of the affect itself (Dolan et al, 2000).

Despite the increasing recognition of GAD as an independent disorder, it remains important to emphasize that GAD is often followed by depression. Several kinds of explanation have been give to explain this temporal relationship (Table 3.2). In patients with GAD–depression, symptoms are not restricted to anticipation of future harm.

Neurocircuitry of GAD

The literature on the pathogenesis of GAD remains at an early stage. Nevertheless, a number of themes have emerged. A first question is whether the pathogenesis of GAD differs in any way from that of depression. An influential twin study indicated that GAD and major depression (MD) shared common genetic factors, but had substantially different nonfamilial environment risk factors with different kinds of life events predisposing to anxiety and mood disorders (Kendler et al, 1992).

Table 3.2 Explanations of the temporal relationship between GAD and depression.

Biological: dysfunction in gamma-aminobutyric acid (GABA) systems mediates anxiety, and may ultimately lead to changes in monoamine systems and depression (Roy-Byrne and Katon, 1997)
Ethological: after maternal separation, infant primates show protest (a prototype of anxiety) and then later on despair (a prototype of depression) (Bowlby, 1980)
Cognitive: anxiety involves early uncertain helplessness in the face of stressors; depression sets in only after hopelessness becomes apparent (Beck, 1967)

Nevertheless, methodological aspects of this work have been criticized (Kessler et al, 1999), and there is also an empirical literature suggesting that GAD and MD are mediated by separate genetic factors, consistent with the increased acceptance of GAD as an independent disorder. Thus, for example, in another twin study, there was differential transmission of GAD and MD (Torgersen, 1990).

Indeed, preliminary brain imaging studies suggest that GAD is characterized by a number of specific abnormalities (Figure 3.1). Thus, there may be increased amygdala volume (De Bellis et al, 2000b), and abnormal benzodiazepine receptor binding in the temporal pole (Tiihonen et al, 1997b) of GAD patients. An early topographic electroencephalography study indicated differences between GAD and normals in temporal regions (Buchsbaum et al, 1985), and subsequent PET studies have also shown temporal abnormalities in this disorder (Wu et al, 1991).

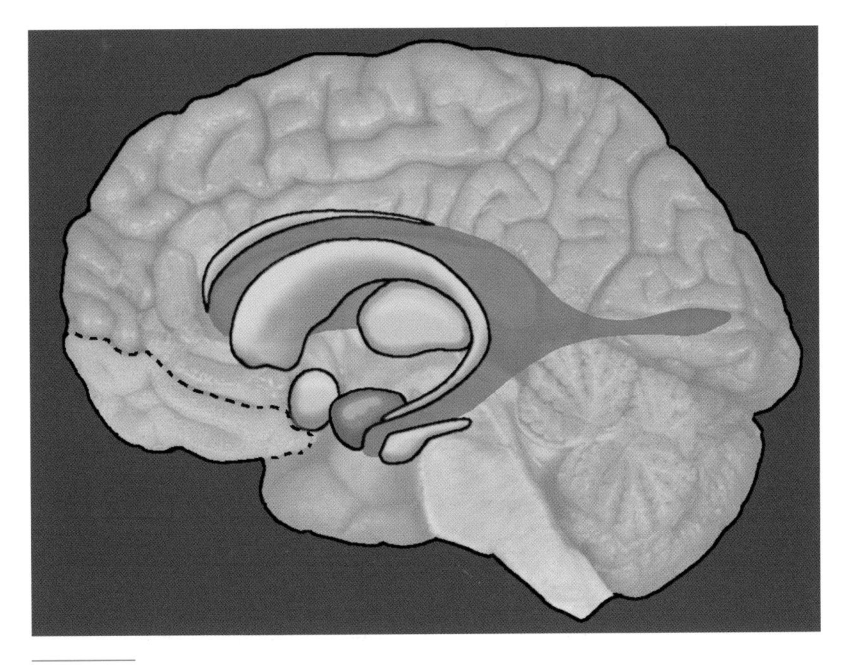

Figure 3.1 *Functional neuroanatomy of GAD: increased activity in amygdala and perhaps prefrontally.*

GAD may also be characterised by increased activity in frontal regions (Wu et al, 1991). Furthermore, an analysis of pooled PET symptom provocation data in obsessive-compulsive disorder (OCD), post-traumatic stress disorder (PTSD), and simple phobia found activation of right inferior frontal cortex, paralimbic structures (right posterior medial orbitofrontal cortex, bilateral insular cortex), bilateral lenticulate nuclei, and bilateral brainstem foci (Rauch et al, 1997b). This finding arguably provides indirect evidence for a role for inferior frontal cortex and the paralimbic system, which serves as a conduit from sensory, motor, and association cortex to the limbic system itself, in a range of anxiety symptoms.

Neurocircuitry and neurochemistry

Animal studies demonstrate that serotonin hypofunction is associated with hypersensitivity to environmental cues and increasing responsiveness to threat (Handley, 1995; Lightowler et al, 1994). Knock-out models reinforce the importance of a link between particular 5-HT subreceptors and anxiety (Parks et al, 1998; Ramboz et al, 1998). Furthermore, in the social interaction test, which may, despite its name, be a model of GAD, SSRIs have anxiolytic effects (Lightowler et al, 1994).

In community studies, there may be an association between the s/s allele of the (5-HTTP) serotonin transporter protein and trait anxiety, but not all studies are consistent (Lesch, 2001). In GAD, however, reduced cerebrospinal fluid (CSF) levels of serotonin and reduced platelet paroxetine binding have been observed (Iny et al, 1994). Furthermore, administration of the non-specific serotonin agonist m-chlorophenylpiperazine (mCPP) resulted in increased anxiety and hostility in GAD patients (Germine et al, 1992).

In addition, treatment data support the possibility that the serotonin system plays a role in mediating GAD. Buspirone, a 5-HT_{1A} partial agonist, has long been used for the treatment of this disorder, despite uncertainty about the robustness of its effects. Moreover, the SSRIs are increasingly viewed as first-line agents in the management of GAD (Ballenger et al, 2001).

There is growing interest in imaging the role of the serotonin system in anxiety (Tauscher et al, 2001). To date, however, there are few available data on the effect of SSRI treatment on the functional neuroanatomy of GAD. Nevertheless, it may be hypothesized that as in other mood/anxiety disorders, these agents are able to normalize dysfunctional circuits (Figure 3.2).

Other systems also play a role in GAD. The noradrenergic system, for example, has been suggested to play a crucial role in focusing attention onto salient events in threatening or demanding situations (Robins and Everitt, 1995). Certainly, the noradrenergic system is strongly implicated in animal models of fear (Redmond, 1986).

In clinical studies of GAD, increased plasma norepinephrine and 3-methoxy-4-hydroxyphenylglycol (MHPG), and reduced platelet α_2-adrenergic peripheral receptor binding sites have been reported,

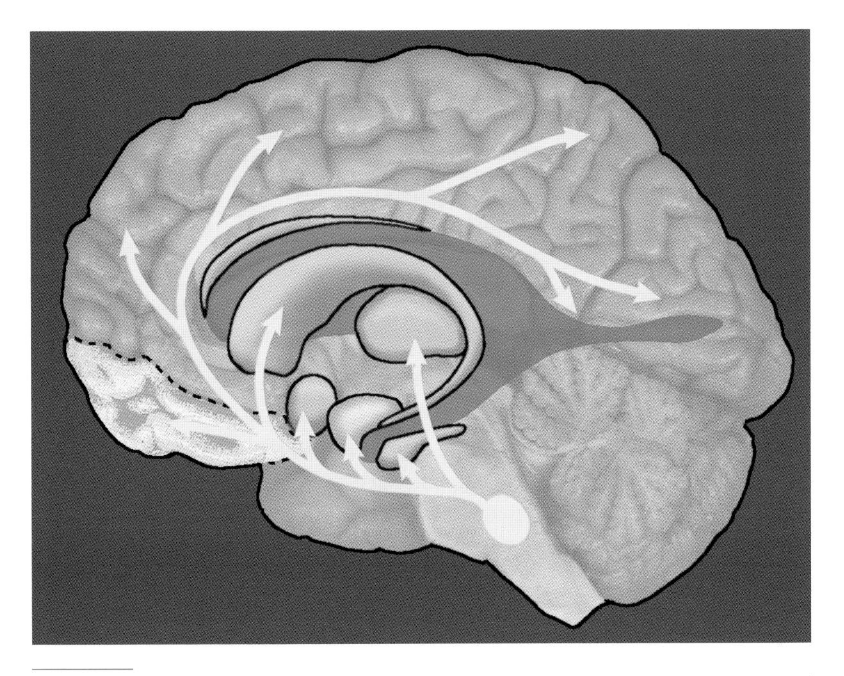

Figure 3.2 *Effects of SSRIs on functional neuroanatomy of GAD: normalization of activity in amygdala and prefrontal region.*

although not all studies of static noradrenergic measures have been consistent. Administration of more dynamic adrenergic probes has indicated reduced adrenergic receptor sensitivity, perhaps an adaptation to high circulating catecholamines (Nutt, 2001).

Furthermore, the dual serotonergic and noradrenergic reuptake inhibitor venlafaxine appears useful in the treatment of GAD (Sussman and Stein, 2002). The locus ceruleus (LC) projects to the amygdala and to other structures involved in the anxiety response, so that noradrenergic involvement is not inconsistent with the neuroanatomic model outlined above.

Involvement of the gamma-aminobutyric acid (GABA) receptor complex in GAD is supported by a number of studies including those demonstrating the responsiveness of this disorder to benzodiazepines (Sussman and Stein, 2002). Anxious subjects (Weizman et al, 1987) and GAD patients (Rocca et al, 1991) demonstrate reduced benzodiazepine binding capacity, with normalization of findings after benzodiazepine treatment. Furthermore, in a study of female GAD patents, there was significant reduction in left temporal pole benzodiazepine receptor binding (Tiihonen et al, 1997b). GABA is the brain's predominant inhibitory neurotransmitter and there is a particularly dense distribution of α_2 $GABA_A$ receptors, on which benzodiazepines act, in limbic and paralimbic areas (Löw et al, 2001).

Evolutionary considerations: future alarm

Anxiety is likely to have had a long evolutionary history (Darwin, 1965; Stein and Bouwer, 1997a), occurring in response to signals of danger and representing a set of response tendencies that have resulted in avoidance of similar dangers in the history of both the individual organism and its species. Different kinds of fear responses tend to emerge at the developmental stage where they become adaptive.

Even in simple organisms, like Aplysia, simple conditioning and sensitization occur. In more complex organisms, the anxiety response includes affective, cognitive and motoric components. Interestingly, there appears to be some continuity in neurobiological mechanisms; thus in Aplysia conditioning results in increased release of serotonin with presynaptic facilitation (Kandel, 1983).

It is important to emphasize that genetic and environmental factors are inextricably intertwined in fear conditioning. Clearly fear conditioning requires learning. However, there has already been progress in delineating the specific genes responsible for determining such factors as increased susceptibility to fear conditioning and/or poor habituation (Flint, 1997).

GAD can be argued to represent a non-specific anxiety/tension alarm (or future alarm). Indeed, it has been argued that general anxiety evolved to deal with threats whose nature could not be defined very clearly, while subtypes of anxiety evolved to deal with particular dangers (Marks and Nesse, 1994). Nevertheless, anxiety defends against many kinds of danger, and subtypes of fear alarms (see later chapters) are not completely distinct.

Argument for an evolutionary-based false alarm can be strengthened if evidence of individuals with a maladaptive low threshold alarm can be found. It turns out that primates who appear to have low harm avoidance and 5-HT hypofunction do have increased morbidity (Higley et al, 1996). Similarly it may be speculated that humans with low thresholds for the anxiety/tension alarm suffer from antisocial personality traits; such people seem not to show adaptive responses to potentially dangerous situations.

Interestingly, early electrophysiological studies of antisocial personality disorder (ASPD) suggested that subjects failed to show 'arousal'. More recently, there has been growing evidence for an association between decreased frontal lobe activity and increased impulsive aggression (Davidson et al, 2001; Stein, 2000). This stands in contrast to increased frontal activity in GAD patients.

Management

Patients with GAD may be experienced as frustrating by their clinicians. Repeated requests for reassurance, difficult to manage somatic symptoms, and escalating use of benzodiazepines, may unfortunately provoke angry countertransferance. As in so many disorders, a first step is therefore to agree with patients on an explanatory model of the condition. Once the patient's model of the disorder is understood, a negotiation can begin, in which the clinician's model of GAD as a medical disorder, characterized by cognitive and somatic symptoms, and responsive to various treatment interventions, can begin.

Benzodiazepines and buspirone have long been used for the treatment of chronic anxiety. Nevertheless, the efficacy, broad spectrum of action, and tolerability of newer generation antidepressants has led to the recommendation that these agents are now the pharmacotherapies of choice for GAD (Ballenger et al, 2001). Venlafaxine and some of the SSRIs, in particular, have been registered for this indication. Whereas benzodiazepines are particularly useful for somatic symptoms, antidepressant agents appear useful for both somatic and psychic symptoms. It is also theoretically possible that early treatment with such agents may prevent the onset of the later comorbid depression that so often characterizes the course of untreated GAD.

For all the major anxiety disorders, cognitive-behavioral psychotherapy is also effective. Patients are encouraged to endure the anxiety, and to reduce avoidance. In the case of GAD, this may mean setting aside a specific worry time, and not allowing oneself to worry at other times. Different anxiety disorders may also be characterized by particular anxiety schemas, with attentional bias to different forms of possible harm. During psychotherapy of GAD, specific techniques may be used to activate schemas revolving around future harm (e.g. discussion of important interpersonal events, role-playing), and a process of schema re-evaluation and restructuring can then be instituted.

There is relatively little work comparing pharmacotherapy and psychotherapy of GAD, or investigating their optimal sequencing. There is also very little work that has investigated the management of GAD in clinical settings (as opposed to research trials). Patients may require different interventions at different time points, for example, short term use of benzodiazepines may be useful during particularly stressful times. Clinical judgment remains key in deciding on how best to approach the management of individual patients; appropriate use of a range of different interventions allows the treatment of GAD to be rewarding and often successful.

Conclusion

GAD is a complex disorder that probably cannot be reduced to a dysfunction in any particular cognitive process or neurotransmitter system, nor simply seen as an understandable response to normal stressors. This chapter has argued that patients with GAD experience a world dominated by anxiety and tension; this may involve a false non-specific future alarm, with a number of neuroanatomical and neurochemical systems involved in the relevant wetware.

In clinical practice, it is important to recognize and treat not only GAD, but also the triad of GAD–depression–somatization (Stein et al, 2001). While this triad of symptoms is commonly found throughout the world, its experience and expression may vary from place to place, and time to time. 'Neurasthenia' for example was a common diagnosis in the United States in the nineteenth century, and remains in wide use in the East.

Fortunately, both pharmacotherapy and psychotherapy are effective in the management of GAD. It seems entirely reasonable to speculate that early diagnosis and intervention may prevent the subsequent development of mood and other comorbid conditions. Primary care practitioners, who likely see the majority of GAD patients, must be trained carefully in the recognition and treatment of this important condition.

chapter 4

Obsessive-compulsive disorder

Symptoms and assessment

Obsessive-compulsive disorder (OCD) was the fourth most common psychiatric disorder in the United States Epidemiological Catchment Area (ECA) study (Robins et al, 1984), and has a lifetime prevalence of 2–3% in many parts of the globe (Weissman et al, 1994). Furthermore, it was the tenth most disabling of all medical disorders in a landmark mortality and morbidity study (Murray and Lopez, 1996). Indeed, OCD has been suggested to cost the economy of a country such as the United States several billion dollars each year (Hollander et al, 1997).

OCD is characterized by obsessions and compulsions (Table 4.1). Obsessions are intrusive thoughts, ideas, or images that increase anxiety, whereas compulsions are repetitive rituals or mental actions performed in response to obsessions in order to decrease anxiety. While patients report a wide range of different kinds of obsessions and compulsions, there is an impressive consistency of themes between patients (Table 4.2) and across cultures (Stein and Rapoport, 1996).

Despite this homogeneity, there have been increasing attempts to further the understanding and treatment of OCD by specifying different subgroups (Table 4.3). There is some evidence, for example, that the various symptoms of OCD can be mapped onto a four-factor solution, with each symptom factor being mediated by somewhat different neurobiological factors (Leckman et al, 1997). Thus, the presence

Table 4.1 Symptoms of OCD (modified from DSM-IV-TR).

(A) Either obsessions or compulsions.

Obsessions as defined by:

(1) recurrent and persistent thoughts, impulses, or images that are experienced, at some time during the disturbance, as intrusive and inappropriate and that cause marked anxiety or distress
(2) the thoughts, impulses, or images are not simply excessive worries about real-life problems
(3) the person attempts to ignore or suppress such thoughts, impulses, or images, or to neutralize them with some other thought or action
(4) the person recognizes that the obsessional thoughts, impulses, or images are a product of his or her own mind (not imposed from without as in thought insertion)

Compulsions as defined by:

(1) repetitive behaviors (e.g. hand washing, ordering, checking) or mental acts (e.g. praying, counting, repeating words silently) that the person feels driven to perform in response to an obsession, or according to rules that must be applied rigidly
(2) the behaviors or mental acts are aimed at preventing or reducing distress or preventing some dreaded event or situation; however, these behaviors or mental acts either are not connected in a realistic way with what they are designed to neutralize or prevent or are clearly excessive

(B) At some point during the course of the disorder, the person has recognized that the obsessions or compulsions are excessive or unreasonable.

(C) The obsessions or compulsions cause marked distress, are time consuming (take more than one hour a day), or significantly interfere with the person's normal routine, occupational (or academic) functioning, or usual social activities or relationships.

(D) If another disorder is present, the content of the obsessions or compulsions is not restricted to it.

(E) The disturbance is not due to the direct physiological effects of a substance (e.g. a drug of abuse, a medication) or a general medical condition.

Table 4.2 Typical obsessions and consequent compulsions in OCD.

Obsessions	*Compulsions*
Contamination concerns	Washing, showering
Pathologic doubt	Checking, praying
Symmetry concerns	Ordering, arranging
Hoarding concerns	Hoarding behaviors

Table 4.3 Subtypes of OCD.

Early onset vs late onset (male:female ratio is higher in early onset)
With tics vs without tics (patients with tics are less likely to respond to SSRIs)
Low vs high neurological soft signs (patients with high soft signs are less likely to respond to SSRIs)

Table 4.4 Putative obsessive-compulsive spectrum disorders.

Body dysmorphic disorder: preoccupation with imagined defects in appearance
Olfactory reference syndrome: preoccupation with imagined body malodor
Hypochondriasis: preoccupation with having a medical disorder, despite appropriate evaluation and reassurance
Trichotillomania: recurrent hair-pulling preceded by increased tension and followed by relief
Tourette's syndrome: motor and vocal tics occurring over a lengthy period of time and beginning before adulthood.

of comorbid tics in OCD appears to have specific treatment implications.

Another nosological debate involves the putative obsessive-compulsive spectrum disorders (Table 4.4). Although controversy remains about which disorders belong to this spectrum, a number of conditions share considerable phenomenological and neurobiological overlap with OCD. In particular, several disorders characterized by intrusive repetitive symptoms show a selective response to serotonin reuptake inhibitors (SRIs) in comparison with noradrenaline reuptake inhibitors (NRIs). These include OCD itself, body dysmorphic disorder, possibly olfactory reference syndrome, hypochondriasis, trichotillomania, and obsessive-compulsive symptoms in Tourette's syndrome (TS) and autism. There may also be overlap in the neuroanatomy, neuroimmunology, and neurogenetics of some of these disorders, particularly between OCD and TS (Stein, 2001b).

The severity of symptoms in OCD and in a number of putative obsessive-compulsive spectrum disorders can be measured using the Yale–Brown Obsessive-Compulsive Scale (YBOCS) (Goodman et al, 1989) (Table A.3, see Appendix). After completing a symptom checklist, patients are administered 5 items addressing obsessions and 5 items addressing compulsions. The scale is user-friendly and there is good evidence for both reliability and validity. A children's version also exists.

Cognitive-affective considerations: procedural strategies

A crucial component of normal cognitive-affective functioning is the selection, maintenance, and initiation of cognitive and motoric programs. These programs have been given terms such as 'habit system' (Mishkin and Petri, 1984), 'response set' (Robins and Brown, 1990), or 'procedural mobilization' (Saint-Cyr et al, 1995). Given that OCD symptoms involve stereotypical behavior, an immediate possibility is that OCD involves dysfunction of procedural strategies.

It is likely that cortical–striatal–thalamic–cortical (CSTC) systems play a crucial role in the implicit learning of procedural strategies and their subsequent automatic execution. There are several parallel CSTC circuits, each of which governs a somewhat different spectrum of cognitive and affective function (Alexander et al, 1986) (Table 4.5, Figures 4.1–4.5).

Ventral CSTC (Figure 4.3) circuits appear to play a particularly import-

Table 4.5 Cortical–striatal–thalamic–cortical (CSTC) circuits.

Circuit	*Frontal cortex*	*Striatum*	*Thalamus*
Sensorimotor	Motor somatosensory	Putamen	Ventral lateral Ventral anterior
Dorsal	Dorsolateral prefrontal	Dorsolateral caudate	Ventral anterior Medial dorsal
Ventral	Lateral orbital cortex	Ventromedial caudate	Ventromedial Medial dorsal
Limbic	Anterior cingulate	Nucleus accumbens	Medial dorsal

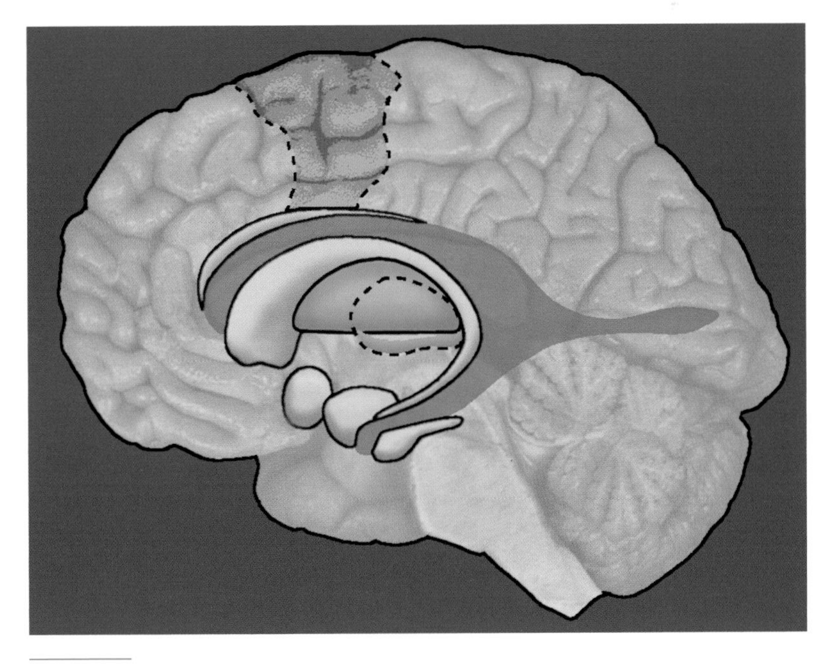

Figure 4.1 *Sensorimotor CSTC circuit: sensorimotor cortex, putamen, thalamus.*

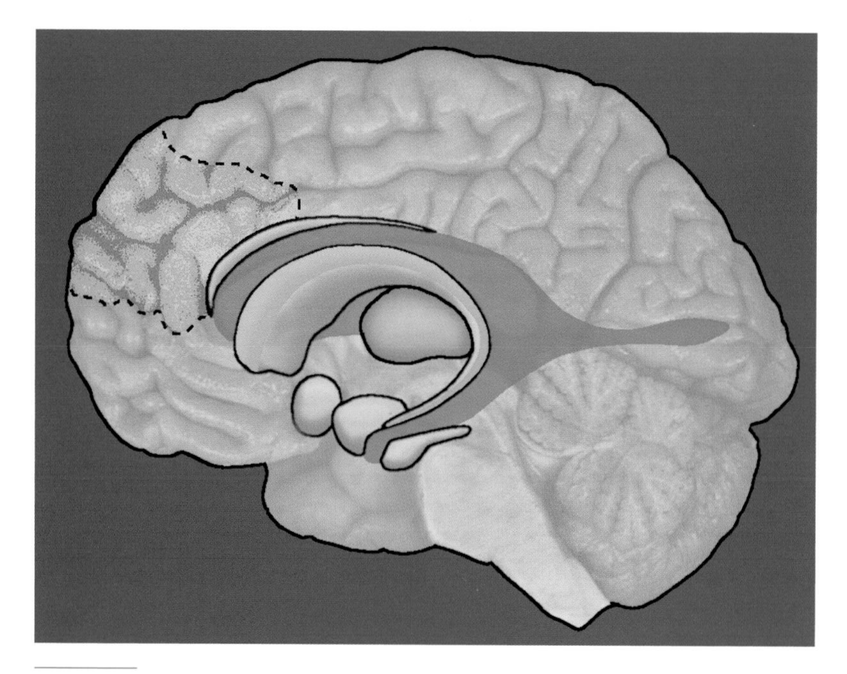

Figure 4.2 *Dorsal CSTC circuit: dorsal frontal cortex, dorsal striatum, thalamus.*

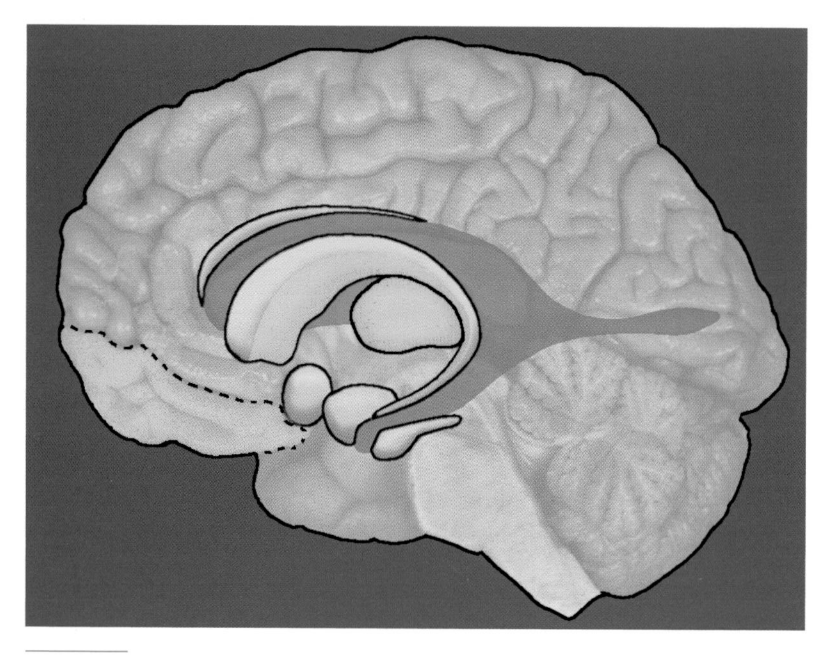

Figure 4.3 *Ventral CSTC circuit: orbitofrontal cortex, ventral striatum, thalamus.*

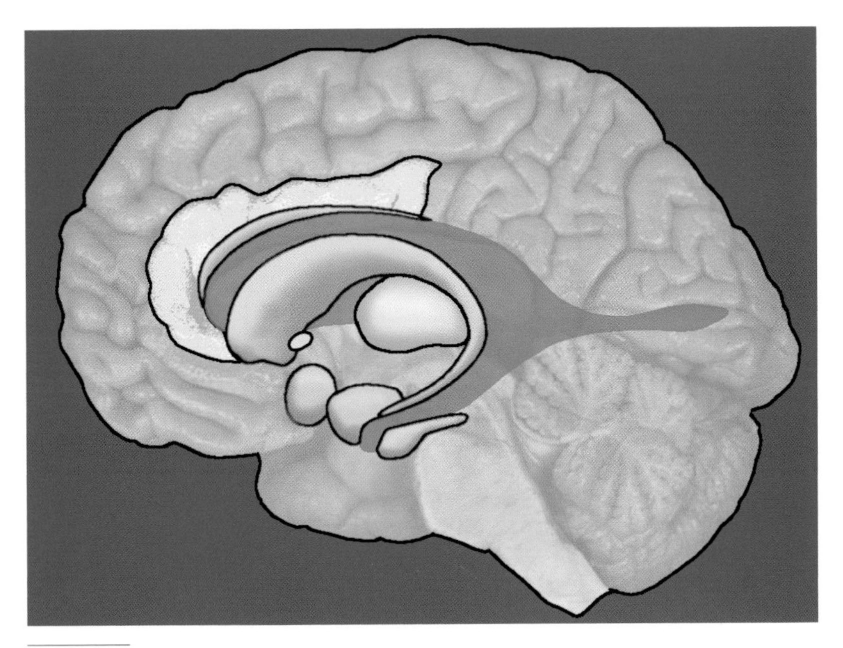

Figure 4.4 *Limbic CSTC: cingulate, nucleus accumbens, thalamus.*

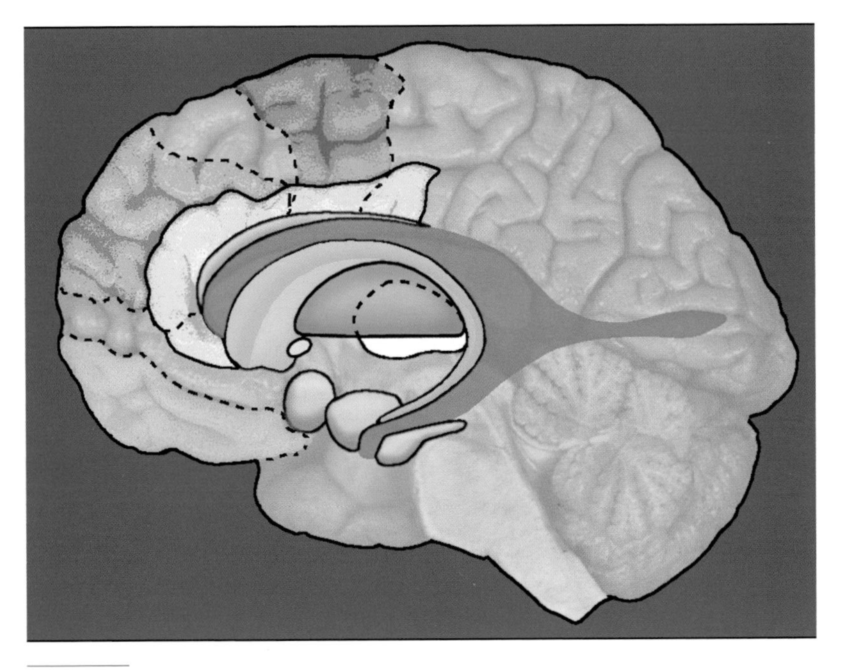

Figure 4.5 *Range of CSTC circuits: sensorimotor, dorsal, ventral, limbic.*

ant role in recognizing behaviorally significant stimuli (and in error detection) and in regulating autonomic and goal-directed responses (including response inhibition and suppression of negative emotion), and are therefore a good candidate for involvement in OCD (Davidson et al, 2001; Rauch and Baxter, 1998; Zald and Kim, 1996). In Tourette's syndrome, more motoric symptoms are mediated by related CSTC circuits (Stern et al, 2000).

Neurocircuitry of OCD

There is a range of evidence that CSTC circuits are disrupted in OCD. Perhaps the first evidence that OCD might have a neurological basis was provided during the pandemic of viral encephalitis lethargica early in the last century. Patients with parkinsonian features were observed to have obsessive-compulsive symptoms, tics, and focal brain lesions including involvement of the basal ganglia (Cheyette and Cummings, 1995).

Other neurological disorders with basal ganglia involvement can also present with obsessive-compulsive symptoms (Stein et al, 1994) (Table 4.6). Of particular interest is the association between streptococcal infection and subsequent OCD and tics (so-called PANDAS, or pediatric autoimmune neuropsychiatric disorders associated with *Streptococcus*), as this association suggests that autoimmune factors may be involved in OCD (Swedo et al, 1998).

A range of additional evidence points to cortico-striatal involvement in OCD (Stein et al, 1994). OCD patients have increased neurological soft

Table 4.6 Lesions of the basal ganglia associated with OCD.

Infectious/immune: postencephalitic parkinsonism, Sydenham's chorea
Traumatic/toxic: Head injury, wasp sting, manganese intoxication
Vascular/hypoxic: Infarction, carbon monoxide intoxication, neonatal hypoxia
Genetic/idiopathic: Tourette's syndrome, Huntington's disease, neuroacanthocytosis

signs including tics, consistent with basal ganglia damage. Neuropsychological testing is arguably also consistent with CSTC dysfunction. Neurosurgery of refractory OCD involves making lesions in CSTC circuitry.

Perhaps most persuasive, however, is evidence from structural and functional brain imaging (Rauch and Baxter, 1998). A number of structural imaging studies have pointed to basal ganglia abnormalities, although there is some inconsistency, with data pointing to increased basal ganglia volume as well as decreased basal ganglia volume. Interestingly, patients with PANDAS have been shown to have larger basal ganglia, with shrinkage perhaps occurring over time; such possible changes in volume over time may contribute to some of the inconsistencies between studies.

A range of functional imaging studies have pointed to CSTC involvement in OCD. Studies show that at baseline OCD patients have hyperactivity in orbitofrontal cortex, anterior cingulate, and caudate nucleus (Figure 4.6). This hyperactivity is exacerbated by exposure to feared

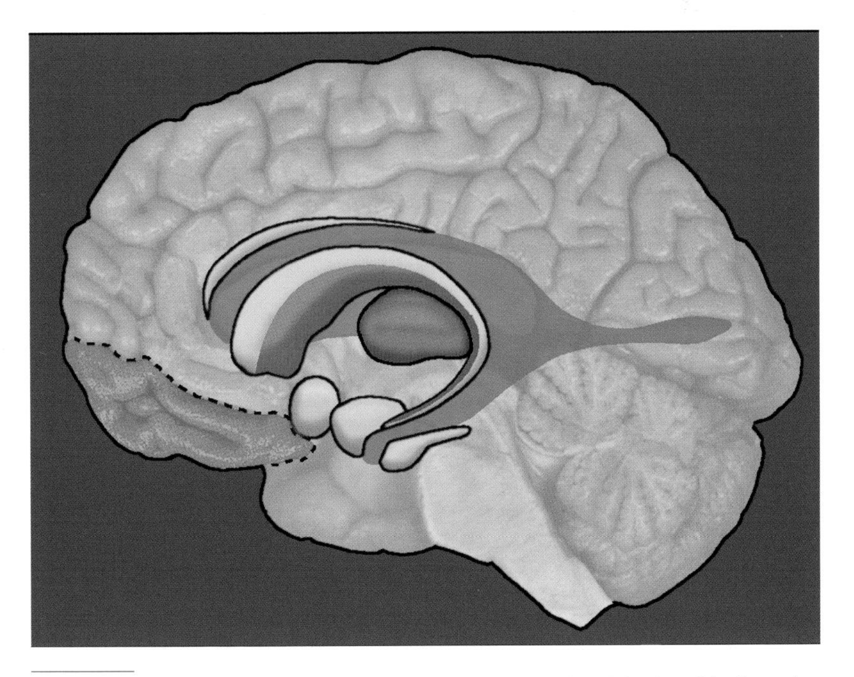

Figure 4.6 *Functional neuroanatomy of OCD: increased activity in orbitofrontal cortex and ventral striatum.*

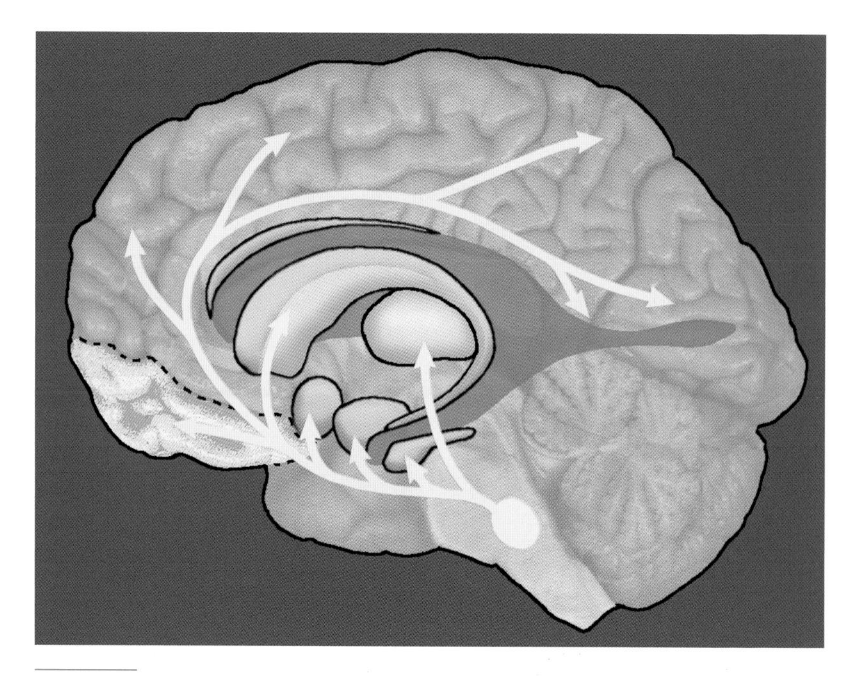

Figure 4.7 *Effects of SSRIs on functional neuroanatomy of OCD: normalization of activity in orbitofrontal and ventral striatal regions.*

stimuli, but is normalized by successful treatment with either medication or behavioral therapy (Baxter et al, 1992) (Figure 4.7).

The question of how CSTC dysfunction arises in OCD remains unresolved. A range of mechanisms may be involved, including genetic factors, the marked sensitivity of striatal circuits to anoxic damage, the development of disordered striatal architecture after emotional deprivation, as well as a broad spectrum of potential neurochemical, neuroimmunological, and neuroendocrinological factors.

Neurocircuitry and neurochemistry

As noted earlier, OCD is well known to have a selective response to serotonergic agents. Several other findings support the role of serotonin in mediating OCD. The literature on peripheral markers in OCD is somewhat inconsistent, but an early study of OCD showed that CSF 5-HIAA decreased during treatment with clomipramine (Thoren et al, 1980). Studies of serotonergic 'challenges' in OCD also show some inconsistency, but there appears to be a group of patients who respond to serotonin agonists with exacerbation of OCD symptoms, a phenomenon that is no longer apparent after SRI treatment (Zohar et al, 1988).

The serotonin system is a complex one, and an aim of recent research has been to focus on the serotonin subreceptors most relevant to OCD. Animal work has, for example, focused on the 5-HT_{1D} receptor; during administration of SSRIs this terminal autoreceptor (i.e. brake) is gradually desensitized, so resulting in increased serotonergic acitivity (i.e. the brake is lifted). In animals, 5-HT_{1D} desensitization requires at least 8 weeks of high-dose SRI treatment, so paralleling clinical findings which indicate that OCD requires higher doses and longer duration of treatment than does depression (El Mansari et al, 1995) (Table 4.7). 5-HT_2 receptors may also be important (Delgado and Moreno, 1998).

Do these findings fit with our knowledge of the neurocircuits involved in OCD? Importantly, CSTC circuits have significant serotonergic innervation. In animal work, 5-HT_{1D} receptors desensitized by relatively high doses and durations of SSRIs lie in orbitofrontal cortex. In humans, particular 5-HT_{1D} alleles have been associated with increased risk for OCD (Mundo et al, 2000), and administration of sumatriptan, a 5-HT_{1D} agonist, is associated

Table 4.7 Treatment principles in depression and OCD.

Depression	*OCD*
Responds to both serotonin and noradrenergic reuptake inhibitors	Responds selectively to serotonin reuptake inhibitors
Adequate clinical trial is 4–6 weeks	Adequate clinical trial is 12 weeks
May respond at relatively low doses of a serotonin reuptake inhibitor	May require relatively high doses of a serotonin reuptake inhibitor
Lithium and thyroid hormone may be used to augment treatment response	Dopamine blockers may be used to augment treatment response

with changes in the activation of cortico-striatal circuits in OCD (Stein, 1999). Most persuasively, after treatment of OCD with SSRIs, there is normalization of cortico-striatal activity (Baxter et al, 1992) (Figure 4.7).

Systems other than serotonin are also likely to play a crucial role in mediating OCD. After all, only around 40–60% of patients respond to SRIs, and only a similar percentage of patients demonstrate abnormal responses to 'challenges' with the serotonin agonist mCPP (Barr et al, 1992). Indeed, it remains quite possible that serotonergic dysfunction per se plays no role, or only a minor role, in the pathogenesis of OCD.

The dopamine system may be particularly important. Administration of dopamine agonists results in increased stereotypies in animals, and may result in obsessive-compulsive symptoms or tics in humans (Goodman et al, 1990). Conversely, dopamine blockers are useful in the treatment of tics, and while OCD patients with tics are less likely to respond to SSRIs, they may well show a response to augmentation with dopamine blockers (McDougle et al, 1994).

Given the dopaminergic innervation of the striatum, and the interaction between the serotonin and dopaminergic systems, these findings are consistent with a CSTC model. Infusion of dopamine into the caudate results in stereotypic orofacial behaviors (grooming, gnawing) in animals (Fog and Pakkenberg, 1971). Dopaminergic striatal circuits are perhaps particularly likely to be important in OCD patients with tics, and in patients with OCD spectrum disorders (such as TS) that are characterized by involuntary movements.

Other systems that may well be involved in OCD include the glutamatergic and opioid systems, certain neuropeptides such as oxytocin, and perhaps even hormonal steroids (Leckman et al, 1994; Stein, 1996). Nevertheless, to date, pharmacological interventions for OCD remain restricted to those that target the dopamine and serotonin systems.

Evolutionary considerations: grooming alarms

OCD can be viewed in terms of a failure to inhibit cortico-striatally governed procedural strategies from intruding into consciousness. Such a view is consistent with (i) the limited number of symptom themes in OCD (e.g. washing, hoarding), and their apparent evolutionary importance; (ii) dysfunction of CSTC circuits in a range of studies, with activation of temporal rather than striatal areas during implicit cognition (Rauch et al, 1997a); and (iii) abnormalities in serotonergic systems which innervate cortico-striatal circuits and which may be associated with disinhibition.

It is interesting that a range of procedural strategies may be relevant to OCD. Much of the literature has focused on contamination and increased grooming. Some animal models appear highly relevant to such a formulation, demonstrating a remarkably similar pharmacotherapy response profile with OCD (Rapoport et al, 1992; Stein et al, 1992). Indeed, some OCD spectrum disorders can perhaps be conceptualized as grooming disorders (Stein et al, 1999a). However, other procedural strategies may also be relevant; these include hoarding (Stein et al, 1999b) and symmetry assessment (which again appears to be mediated by specific evolutionary mechanisms, and which may be particularly pertinent to body dysmorphic disorder).

An important theoretical question in the study of OCD is whether patients are compulsive or impulsive. Freud early on argued that OCD was ultimately a disorder of increased aggression, with compensatory defense mechanisms (Stein and Stone, 1997). Put in more modern terms, OCD can be understood to involve loss of inhibitory mechanisms (perhaps primarily serotonergic), with apparent compensatory factors (for example, increased activity in orbitofrontal cortex, with some evidence of hyperserotonergic activity). Whereas impulsive patients have decreased hypofrontal/serotonin function, compulsive patients are able to also show a compensatory response with hyperfrontal/hyperserotonergic function also (Stein and Hollander, 1993).

Another significant issue is whether OCD is really an anxiety disorder (Montgomery, 1993). Certainly, the primary emotion in OCD does not appear to be fear. Indeed, some authors have begun to suggest rather that the emotion that is particularly relevant to OCD is that of disgust (Stein et al, 2001). Interestingly, the neurobiology of fear and disgust can be dissociated on neurobiological studies; whereas fear involves the amygdala (see Chapter 5), disgust is mediated by the CSTC circuits which are so central to OCD.

Management

A first step in the management of OCD is psychoeducation; assessing the patient's model of their symptoms, and negotiating a shared view. People with OCD often feel relieved to know that many others suffer from very similar shame-filled or senseless obsessions and compulsions. While some view their symptoms in terms of unresolved guilt or similar constructs, many have their own theories about a neuronal 'short-circuit', and it can be satisfying for them to learn that modern science supports a roughly analogous view.

The first-line pharmacotherapy of OCD is a serotonin reuptake inhibitor. Given the superior tolerability of the selective serotonin re-uptake inhibitors, one of these agents is often used first. Nevertheless, meta-analyses of OCD tend to show that clomipramine is a particularly effective agent (Stein, Spadaccini, Hollander, 1995), and this agent is well worth considering in more refractory patients. As discussed earlier, dosages and durations of the SRIs may need to be higher in the treatment of OCD than in depression.

The first-line psychotherapy of OCD is cognitive-behavioural therapy (CBT). Patients with OCD often anticipate that exposure will be difficult if not impossible. Work demonstrating the ability of both pharmacotherapy and psychotherapy to normalize brain activity (Baxter et al, 1992) provides a useful rationale for persuading patients to at least attempt to apply the principles of CBT. Over time, they often become increasingly confident in their ability to fight against the symptoms of OCD.

As in the case of a number of other anxiety disorders, there is relatively little evidence about the optimal combination and sequencing of pharmacotherapy and psychotherapy of OCD. Clinical judgment is required in assessing the best choice of intervention for a particular patient, and incorporation of both modalities is often useful. For example, some patients may require medication before they are able to tolerate CBT. In other patients, CBT may be particularly useful during medication tapering.

Treatment-refractoriness unfortunately remains a problem for a significant proportion of patients with OCD. A range of pharmacotherapy augmentation strategies have been studied, but current evidence most strongly supports the use of low doses of dopamine blockers (McDougle et al, 2000). Given their relatively safety and tolerability, the new generation of antipsychotic medications are increasingly recommended for this purpose. In extremely refractory patients, more invasive interventions (e.g. deep brain stimulation or neurosurgery) can be considered.

Conclusion

There have been very great changes in the OCD field; once conceptualized as a rather rare and treatment-refractory condition, OCD is now recognized to be one of the most common of the psychiatric disorders, and it often responds to modern treatments. Whereas OCD once provided a key exemplar for a model of psychopathology based on unconscious conflict, it can now be seen as a seminal exemplar for a psychiatry based on modern cognitive-affective neuroscience.

OCD is useful for demonstrating a grounded approach to the brain–mind, in so far as it was the first disorder for which it was shown that both specific medications (SRIs) and particular psychotherapeutic techniques (exposure and response prevention) normalized precise neuroanatomical circuitry. It is also a disorder for which an evolutionary approach makes not only theoretical sense given the nature of OCD symptoms (grooming, hoarding, etc.), but for which there are also empirical data showing an overlap in pharmacotherapeutic responses in animals and humans.

In understanding OCD it is readily possible to avoid reductionist approaches which view the disorder merely in terms of neurotransmitter dysfunction (this does not make sufficient sense of the disorder), or which see OCD as a meaningful response to stressors (OCD clearly involves brain–mind dysfunction). Instead, an approach that views OCD as characterized by a false alarm that is embodied in the brain–mind's wetware and that can be triggered by a range of social cues is better able to encompass clinical experience and the available research data.

chapter 5

Panic disorder

Symptoms and assessment

Panic disorder is present in approximately 2% of the population, with a somewhat higher incidence in females (Kessler et al, 1994). Patients frequently present to primary care practitioners and non-psychiatric medical specialists, and underdiagnosis/undertreatment and overutilization of medical resources remain important issues.

Panic attacks may be present in all of the anxiety disorders. However, in panic disorder they are characteristically spontaneous. They are accompanied by a range of symptoms, including respiratory, cardiovascular, gastrointestinal, and occulovestibular symptoms (Tables 5.1 and 5.2). Panic attacks vary, however, in their cueing, in their extent, and in the time of onset (Table 5.3).

Patients may go on to develop agoraphobia, or avoidance of situations which may precipitate panic attacks. This sequence of anxiety-avoidance is a common theme throughout this volume. From a theoretical perspective it raises questions about the different neurocircuits involved in mediating these phenomena; from a clinical perspective it emphasizes the importance of early exposure in preventing later avoidance.

In addition, panic disorder may be accompanied by a number of mood and anxiety disorders, with depression a particularly important complication. Panic–depression is the most common form of mood–anxiety comorbidity (Roy-Byrne et al, 2000), and a number of

Table 5.1 Symptoms of panic attack (modified from DSM-IV-TR).

A discrete period of intense fear or discomfort, in which four (or more) of the following symptoms developed abruptly and reached a peak within 10 minutes:

- palpitations, pounding heart, or accelerated heart rate
- sweating
- trembling or shaking
- sensations of shortness of breath or smothering
- feeling of choking
- chest pain or discomfort
- nausea or abdominal distress
- feeling dizzy, unsteady, lightheaded, or faint
- derealization (feelings of unreality) or depersonalization (being detached from oneself)
- fear of losing control or going crazy
- fear of dying
- paresthesias (numbness or tingling sensations)
- chills or hot flushes

authors have emphasized the importance of the link between panic and suicide (although not all data are consistent).

Panic attacks can be measured by a number of different scales, such as the Panic and Agoraphobia Scale (Bandelow, 1998) (Table A.4, see Appendix), which assesses both panic symptoms and avoidance behaviors. Patients should also receive a careful medical history and examination in order to rule out general medical conditions that can present with anxiety symptoms.

Cognitive-affective considerations: conditioned fear

An important recent advance in cognitive-affective neuroscience has been the delineation of brain circuits involved in fear conditioning. Fear

Table 5.2 Symptoms of panic disorder (modified from DSM-IV-TR).

(A) Both (1) and (2):
- (1) recurrent unexpected panic attacks
- (2) at least one of the attacks has been followed by 1 month (or more) of one (or more) of the following:
  - (a) persistent concern about having additional attacks
  - (b) worry about the implications of the attack or its consequences (e.g. losing control, having a heart attack, 'going crazy')
  - (c) a significant change in behavior related to the attacks

(B) The panic attacks are not due to the direct physiological effects of a substance (e.g. a drug of abuse, a medication) or a general medical condition (e.g. hyperthyroidism).

(C) The panic attacks are not better accounted for by another mental disorder, such as social phobia (e.g. occurring on exposure to feared social situations), specific phobia (e.g. on exposure to a specific phobic situation), obsessive-compulsive disorder (e.g. on exposure to dirt in someone with an obsession about contamination), post-traumatic stress disorder (e.g. in response to stimuli associated with a severe stressor), or separation anxiety disorder (e.g. in response to being away from home or close relatives).

Table 5.3 Panic attack subtypes in panic disorder.

Spontaneous vs cued
Full vs partial
Daytime vs nocturnal

conditioning has been a fundamental paradigm for clinicians since the eminent behaviorist John Watson demonstrated how a fear of fluffy white toys could be conditioned in an infant by pairing presentation of such toys with an aversive stimulus. Modern neurobiologists have been able to demonstrate the proximate neurobiological factors involved (Davis and Whalen, 2001; Le Doux, 1998).

The amygdala appears to play a particularly important role in mediating conditioned fear. Afferents to the basolateral amygdala include the thalamus, which relays sensory information; while efferents from the central amygdala nucleus and lateral bed nucleus of the stria terminalis (BNST) (or extended amygdala) include a range of structures, which mediate the 'fight-or-flight' response. Such structures include the lateral nucleus (autonomic arousal and sympathetic discharge) and paraventricular nucleus (increased adrenocorticoid release) of the hypothalamus, as well as the locus ceruleus (increased noradrenaline release), parabrachial nucleus (increased respiratory rate), periaqueductal gray (defensive behaviors and postural freezing), nucleus pontine reticularis (startle response) in the brainstem, and the facial motor nerve (facial expression of fear). Given this information, it is easy to speculate that the amygdala is important in mediating panic attack symptoms.

The hippocampus plays an important role in processing the context (including spatial aspects) of the fear conditioning. The hippocampus is situated at the confluence of a dorsal pathway for spatial/perceptual memory (of 'where') and a ventral pathway for object/conceptual memory (of 'what'), for representing spatial location within a framework fixed to the environment (Burgess et al, 1999). The hippocampus may play a particularly important role in mediating avoidance behaviors in people who have experienced panic attacks.

Consistent with evidence that implicit and explicit processing are partially localized to somewhat different neuroanatomical circuits (Salloway et al, 1997) (Figure 5.1 and Table 5.4). The explicit memory of where and how fear conditioning took place, and the implicit processing involved in the fear conditioning itself, are processes that can be dissociated. Thus, an amygdala lesion does not impact on explicit recall, while a hippocampal lesion does not prevent implicit fear conditioning from occurring (Bechara et al, 1995). There is evidence that the implicit pathway is evolutionarily older (Reber, 1993); certainly the development of explicit memory is likely to have appeared relatively late in the development of primates. Conversely, when there are widespread brain lesions (for example, in Alzheimer's disorder), explicit processing is affected first, while implicit memory is more robust and is retained for longer.

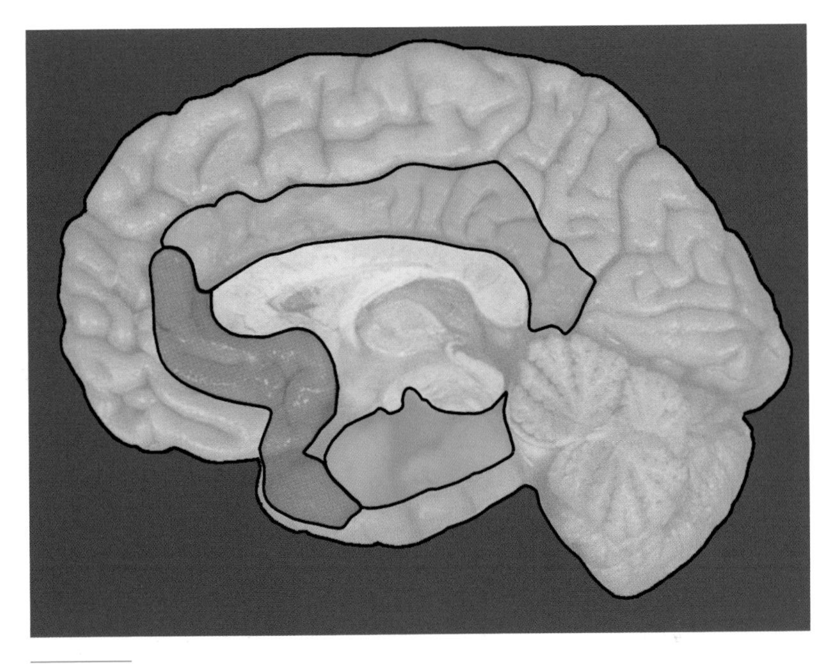

Figure 5.1 *Implicit (red) vs explicit (green) neuronal circuits.* (Adapted from Salloway S, Malloy PF and Cummings JL (eds): The Neuropsychiatry of Limbic and Subcortical Disorders, Washington, DC: American Psychiatric Press, 1998. www.appi.org. Used with permission.)

Table 5.4 Paralimbic circuits.

	Orbitofrontal	*Parahippocampal*
Structures	Amygdala	Hippocampus
Other structures	Infracallosal cingulate Anterior parahippocampus Insula/temporal pole	Supracallosal/posterior cingulate Posterior parahippocampus Retrosplenium
Function	Implicit processing	Explicit processing

Neurocircuitry of panic disorder

Is there any evidence that panic disorder involves dysregulation of amygdala–hippocampal fear systems? Certainly, there is a small literature demonstrating that panic disorder can be associated with a range of (especially right-sided) temporal abnormalities including seizure disorder (Young et al, 1995). Similarly, stimulation of the amygdala in preclinical and clinical studies (of seizure disorder patients) is associated with fear responses (Cendes et al, 1994; Davis and Whalen, 2001). Conversely, patients with amygdala lesions demonstrate selective impairment in the recognition of fearful facial expressions, and show an inability to be conditioned to fear – this is the classical Klüver–Bucy syndrome (Klüver and Bucy, 1939).

Studies in normals show activation of amygdala and periamygdaloid cortical areas during fear acquisition and extinction (Gorman et al, 2000), although this may be related to affective processing rather than affect itself (Davis and Whalen, 2001). A structural imaging study suggested abnormal temporal lobe volume in panic (Vythilingam et al, 2000). In addition, PET scanning during anxious anticipation in normals (Reiman et al, 1989a), and during lactate-induced panic attacks in panic disorder patients (Reiman et al, 1989b) demonstrated increased activity in paralimbic regions (temporal poles) (Figure 5.2). Activation of the amygdala leads in turn to activation of different circuits (hypothalamus, brainstem) (Figure 5.3).

An early study suggested that only panic disorder patients susceptible to lactate-induced panic had abnormal asymmetry of a parahippocampal region at rest (Reiman et al, 1986). Subsequent functional imaging studies have confirmed dysfunctions of hippocampus or parahippocampal regions in panic disorder, although the precise abnormalities documented have not always been consistent (Gorman et al, 2000).

Hypocapnia-induced vasoconstriction compounds the difficulty of interpreting some panic disorder imaging studies in which there is decreased activity in various regions. Another possibility, however, is that whereas anxiety may be associated with activation of particular brain regions in an attempt to suppress negative emotion, during the height of a panic attack there are in fact regions of deactivation (perhaps

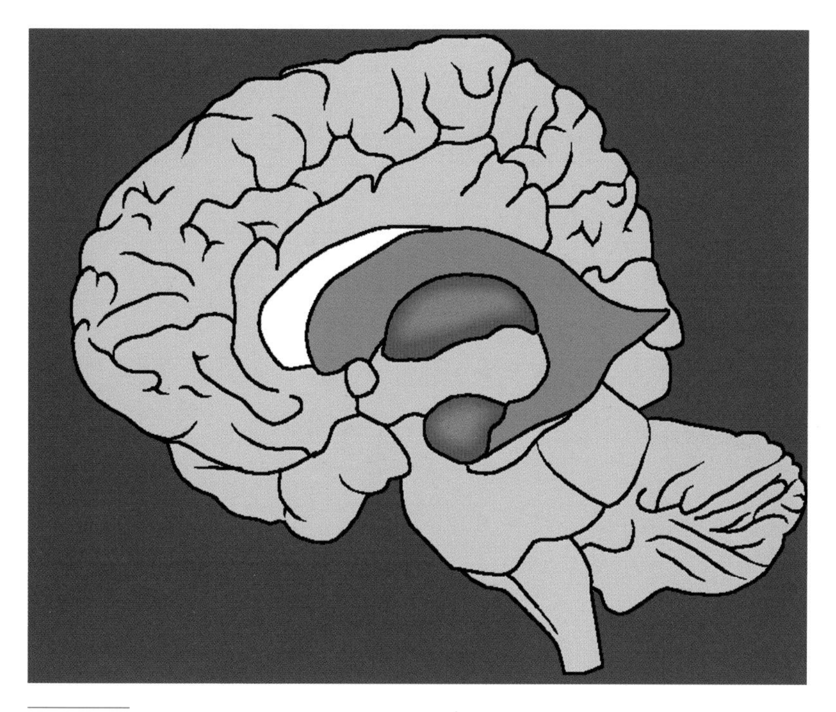

Figure 5.2 *Rapid thalamo-amygdala pathways may be activated during panic attacks.*

associated with impairment in specific cognitive-affective phenomena, such as articulation/verbalization of feelings).

Neurocircuitry and neurochemistry

Several studies of serotonergic markers have shown abnormalities in panic disorder. mCPP, a serotonin agonist, results in exacerbation of panic symptoms in patients with panic disorder. Similarly, panic attacks can be precipitated or exacerbated by a range of substances, such as marijuana, that have serotonin agonist effects (Coplan et al, 1992). One possibility is that postsynaptic serotonin receptors are upregulated in this condition, so that they are supersensitive to such serotonergic agents.

Conversely, there is now good evidence of the efficacy of SSRIs

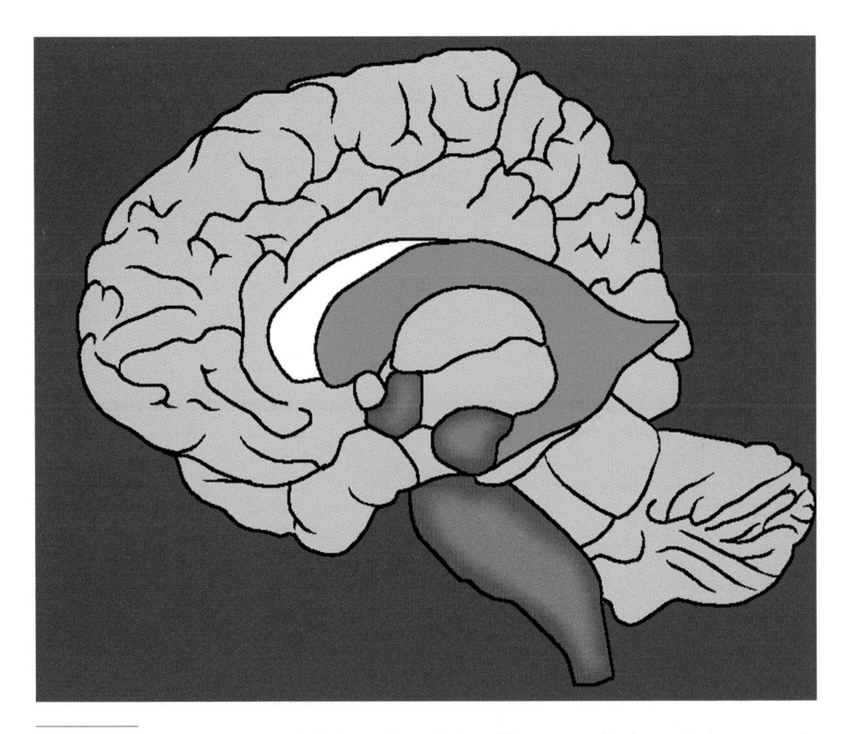

Figure 5.3 *Once the amygdala is activated, its efferents to the hypothalamus and brainstem mediate a fear response.*

in panic disorder, so that these agents are generally considered to be a first-line medication for the treatment of this condition (provided that relatively low doses are used at first, so as to avoid unnecessary agitation). Indeed, an early meta-analysis comparing SSRIs with imipramine and benzodiazepines suggested superiority of this class of agents over other medication (Boyer, 1995). Subsequent effect sizes in SSRI studies have, however, not been as strong.

Is it possible to integrate this data with the work on the neuroanatomy of fear conditioning? It turns out that the serotonergic system interacts at several points with fear conditioning pathways (Coplan and Lydiard, 1998), so allowing an interweaving of neuroanatomic and neurochemical models. Serotonergic projections from the dorsal raphe nucleus (DRN) generally inhibit the locus ceruleus (LC), while projections from the LC stimulate DRN serotonergic neurons and inhibit median raphe nucleus (MRN) neurons. In addition, the DRN sends projections to prefrontal cortex, amygdala, hypothalamus, and periaqeductal gray amongst other structures.

Modulation of the serotonin system therefore has the potential to influence the major regions of the panic disorder circuit, so resulting in decreased noradrenergic activity, diminished release of corticotropin release factor, and modification of defense/escape behaviors. While this kind of model requires additional empirical validation, an interesting imaging study found that after administration of the serotonin releaser and reuptake inhibitor fenfluramine, panic disorder patients had increased parietal-temporal cortex activation (Meyer et al, 2000). Presumably SSRIs are able to normalize functional abnormalities in panic disorder (Bell et al, 2002).

On the one hand, a wide range of agents (other than serotonergic agents) are able to act as panicogenics in patients with panic disorder, consistent with the hypothesis that the amygdala and its efferents serve as a final common pathway for a range of biological stressors (Gorman et al, 2000). On the other hand, certain panicogens (such as CO_2) seem to be particularly robust and to elicit attacks that are strongly reminiscent of clinical panic (Papp et al, 1993), suggesting the centrality of specific neurobiological factors and systems in mediating panic attacks.

One system, for example, that may play a particularly important role in mediating panic attacks is the noradrenergic system. From a basic perspective, the LC receives viscerosensory input and sends afferents to the amygdala, hypothalamus, and brainstem periaqueductal gray. Furthermore, there are several clinical studies that are consistent with a role for noradrenaline; thus, for example, administration of yohimbine, a presynaptic α_2-antagonist, resulted in greater increases in MHPG in panic disorder than in healthy controls in some work (Charney et al, 1984).

In animal models, direct administration of a benzodiazepine agonist produces anxiogenic effects, an effect that is weakened by pretreatment with a benzodiazepine receptor antagonist (Coplan and Lydiard, 1998). While the GABA/BZ receptor is widely distributed in the brain, the basolateral and lateral amygdala nucleus and the hippocampus have high densities. Furthermore, there is evidence of reduced GABA concentrations (Goddard et al, 2001) and of reduced hippocampal and precuneus benzodiazepine receptor binding in panic disorder (Bremner et al, 2000b), although not all findings are consistent.

Evolutionary considerations: suffocation alarms

Klein has argued persuasively that panic disorder is characterized by a false suffocation alarm (Klein, 1993). He begins with the suggestion that the suffocation alarm is an evolved, adaptive response to a lack of oxygen (signaled by increasing pCO_2 and brain lactate). He then goes on to hypothesize, supporting his argument with a review of the relevant empirical literature, that the threshold for this alarm is lowered in panic disorder.

First, the most prominent symptom of many panic attacks is that of dyspnea (indicating a specific emergency reaction to suffocation), and a range of studies document respiratory abnormalities in panic (such as increased sighing, perhaps indicating an attempt to avoid dyspnea by lowering pCO_2). This focus also helps bolster Klein's initial differentiation of episodic spontaneous panics (in which chronic hyperventilation acts as a predictor of lactate-induced panic) from chronic fear-like anxiety (where there is rather HPA activation) (Klein, 1964).

Second, panic attacks increase during a range of conditions that are characterized by an increase in pCO_2 (relaxation and sleep, premenstrual period, and respiratory insufficiency). Conversely, panic attacks decrease during conditions characterized by a decrease in pCO_2 (pregnancy). Interestingly, there appears to be a condition in which the false suffocation alarm is absent; in congenital hypoventilation syndrome or Ondine's curse, patients not only require treatment with agents (e.g. amphetamines) that act as panicogens in normals, but it turns out that they rarely develop panic attacks (Pine et al, 1994).

Another possibility, however, is that panic represents an acute danger alarm, that may be triggered by a range of unconditioned stimuli including increased pCO_2. Indeed, Klein's original conceptualization of panic attacks was in terms of an evolved response to separation anxiety (Klein, 1981). Although such a broader view of the panic disorder 'false alarm' may arguably be weakened by some loss of focus on respiratory systems, it may be more consistent with the range of other data about environmental precipitants of panic attacks (Shear, 1996) and the neurobiology of unconditioned fear responses (Panskepp, 1998).

Management

As in the case of other anxiety disorders, psychoeducation is an important principle in the treatment of panic disorder. Patients with panic attacks are often relieved by the information that their symptoms are distressing but not dangerous. Particularly when pharmacotherapy or psychotherapy is initiated, patients need to be reminded of this important principle.

The first-line pharmacotherapy for panic disorder comprises a SSRI or venlafaxine (Ballenger et al, 1998). While older agents, including the benzodiazepines and tricyclics may well be effective, they have significant disadvantages, including potential adverse events and withdrawal problems. It is important to begin medication at doses lower than those used in the treatment of depression, in order to avoid initial side effects.

From a psychotherapy point of view, cognitive-behavioural principles are key. Patients can be taught relaxation techniques in order to cope with the anxiety felt during panic attacks. Decreasing behavioral avoidance is another crucial goal. As patients become more confident in their ability to use CBT techniques, they can be gradually encouraged to bring on panic attacks (by hyperventilation or exercise), in order to practice their coping skills.

Pharmacotherapy and psychotherapy can be combined as indicated. There is some evidence, however, that CBT may be less effective when patients are treated with benzodiazepines rather than antidepressants. It may be well worth revising CBT techniques and exercises prior to tapering medication, so that should symptoms recur the panic disorder patient is able to manage these without necessarily reinstating pharmacotherapy.

There has been suprisingly little attention in the literature to the management of treatment-refractory panic. Given the morbidity of panic disorder, this is certainly an area that requires further investigation. There is a growing literature on the value of switching antidepressants in the management of treatment-refractory depression (Thase et al, 2002), and similar principles may well hold for the management of panic disorder and other anxiety disorders.

Conclusion

Panic disorder has for many years been a 'lost' disorder, hidden in the broad swath of 'anxiety neurosis'. Its discovery as a unique entity, characterized by specific neurobiological dysfunctions, and responding to selective treatment with modern medications and psychotherapies, represents a tremendously important advance for psychiatry.

Delineation of the neurobiology of fear conditioning has been a particularly influential development for the clinical conceptualization of panic disorder; this move is consistent with the focus of this volume on the crucial importance of neurocircuitry to psychopathology, and on the continuity between animal and human, both in terms of proximate (psychobiological) and distal (evolutionary) mechanisms.

chapter 6

Post-traumatic stress disorder

Symptoms and assessment

Post-traumatic stress disorder (PTSD) has long been considered a 'normal' response to an 'abnormal' event. However, it turns out that rates of trauma are extremely high, and that only a small percentage of people go on to develop chronic PTSD. Thus PTSD is increasingly seen as an abnormal response to a traumatic event (Yehuda and McFarlane, 1995). Certainly, PTSD is a disorder that is characterised by considerable distress and functional impairment.

By definition, then, PTSD begins in the aftermath of a traumatic experience (Table 6.1). DSM-IV attempts to define the traumatic event in both objective terms (there is physical danger) and subjective ones (there is horror, fear). Traumatic events classically associated with PTSD include interpersonal traumas such as combat (more common in men) and rape (more common in women), as well as natural disasters. The more severe the trauma, the more likely the person is to develop PTSD.

Three characteristic clusters of PTSD symptoms emerge after the trauma: re-experiencing, avoidance/numbing, and hyperarousal (Table 6.1). Re-experiencing and hyperarousal symptoms are similar in some ways to the 'positive' or panicky symptoms seen in various anxiety disorders, although they are distinguished by their focus on a traumatic event. Avoidance and numbing symptoms are redolent of the various 'negative' or avoidance symptoms that also cut across the anxiety

Table 6.1 Symptoms of PTSD (modified from DSM-IV-TR).

(A) The person has been exposed to a traumatic event in which both of the following were present:

- (1) the person experienced, witnessed, or was confronted with an event or events that involved actual or threatened death or serious injury, or a threat to the physical integrity of self or others
- (2) the person's response involved intense fear, helplessness, or horror

(B) The traumatic event is persistently re-experienced in one (or more) of the following ways:

- (1) recurrent and intrusive distressing recollections of the event, including images, thoughts, or perceptions
- (2) recurrent distressing dreams of the event. Note that in children, there may be frightening dreams without recognizable content
- (3) acting or feeling as if the traumatic event were recurring (includes a sense of reliving the experience, illusions, hallucinations, and dissociative flashback episodes, including those that occur on awakening or when intoxicated)
- (4) intense psychological distress at exposure to internal or external cues that symbolize or resemble an aspect of the traumatic event
- (5) physiological reactivity on exposure to internal or external cues that symbolize or resemble an aspect of the traumatic event

(C) Persistent avoidance of stimuli associated with the trauma and numbing of general responsiveness (not present before the trauma), as indicated by three (or more) of the following:

- (1) efforts to avoid thoughts, feelings, or conversations associated with the trauma
- (2) efforts to avoid activities, places, or people that arouse recollections of the trauma
- (3) inability to recall an important aspect of the trauma
- (4) markedly diminished interest or participation in significant activities
- (5) feeling of detachment or estrangement from others
- (6) restricted range of affect (e.g. unable to have loving feelings)
- (7) sense of a foreshortened future (e.g. does not expect to have a career, marriage, children, or a normal life span)

Table 6.1 *continued.*

(D) Persistent symptoms of increased arousal (not present before the trauma), as indicated by two (or more) of the following:

- (1) difficulty falling or staying asleep
- (2) irritability or outbursts of anger
- (3) difficulty concentrating
- (4) hypervigilance
- (5) exaggerated startle response

(E) Duration of the disturbance (symptoms in criteria (B), (C), and (D)) is more than 1 month.

(F) The disturbance causes clinically significant distress or impairment in social, occupational, or other important areas of functioning.

disorders, although loss of memory (of the traumatic event) is perhaps particularly distinguishing. A range of associated symptoms, including guilt, shame and anger, may play a particularly important role. Furthermore, there is often significant comorbidity of PTSD with other disorders, particularly mood, anxiety and substance use disorders. Conversely, it should be recognized that trauma may be associated with a range of mood and anxiety disorders other than PTSD. Therefore, screening for past traumas should be included in the assessment of all patients who suffer from these conditions.

Symptoms of PTSD can be assessed using the Clinician Administered PTSD Scale (CAPS) or the briefer TOP-8 (Connor and Davidson, 1999) (Table A.5, see Appendix). The CAPS is based on the DSM criteria for PTSD, measuring both the frequency and intensity of each item. The TOP-8 is comprised of only eight items; these were chosen on the basis of their showing a particularly robust response during medication treatment.

Cognitive-affective considerations: going 'off-line'

The amygdala plays a crucial role in an organism's response to danger. Rapid thalamo-amygdala circuits immediately pass sensory information to the central nucleus of the amygdala (CeN), which then coordinates a complex response (see Chapter 5). Efferent fibers from the CeN and lateral bed nucleus of the stria terminalis (BNST) innervate a number of structures which mediate this complex response. These include the lateral nucleus (autonomic arousal and sympathetic discharge) and paraventricular nucleus (increased adrenocorticoid release) of the hypothalamus, as well as the locus ceruleus (increased noradrenaline release), parabrachial nucleus (increased respiratory rate), and periaqueductal gray (defensive behaviors and postural freezing) in the brainstem, and the facial motor nerve (facial expression of fear).

The hippocampus plays an important role in processing the context of the fear conditioning. As discussed in Chapter 5, the explicit memory of where and how fear conditioning took place, and the implicit processing involved in the fear conditioning itself, are processes that can be dissociated. Indeed, what is remarkable about PTSD is the extent to which explicit cognition is taken 'off-line', and to which memory can be thought of as stored in a 'sensorimotor' form rather than in a 'narrative' one. (Notably, PTSD can develop even when head trauma results in loss of explicit memories (Macmillian, 1991).) While such dissociation may be adaptive at the time of the trauma, it may interfere with processing of the traumatic event and subsequent adaptive responses (Brewin, 2001).

The literature concerning animals suggests that fear conditioning can be extinguished by medial prefrontal cortex (anterior cingulate) (Le Doux, 1998), and there is some supportive human imaging data (Davidson et al, 2001; Hugdahl, 1998). From a different perspective, this 'top-down' control can be understood in terms of the 'processing' of the traumatic event. Implicit processes are integrated together with explicit ones, the traumatic event is articulated and integrated with the rest of the person's schemas, sensorimotor memory is augmented with narrative memory, and the person readjusts and adapts. It is possible, however, that a repeated traumatic event will trigger the rapid amygdalo-thalamic fibers, overriding this frontal processing, and resulting in a return of symptoms (the return of the repressed!). Certainly there is a growing literature documenting the long-lasting psychobiological impact of early developmental trauma and of repeated exposure to stressors (Maier, 2001; Sanchez et al, 2001).

Neurocircuitry of PTSD

Brain imaging findings have provided some empirical evidence that such a model of PTSD is in fact at least partially correct (Figure 6.1). Structural findings have focused on decreased hippocampus volume (Rauch et al, 1998). Although not all studies have been consistent (Bonne et al, 2001),

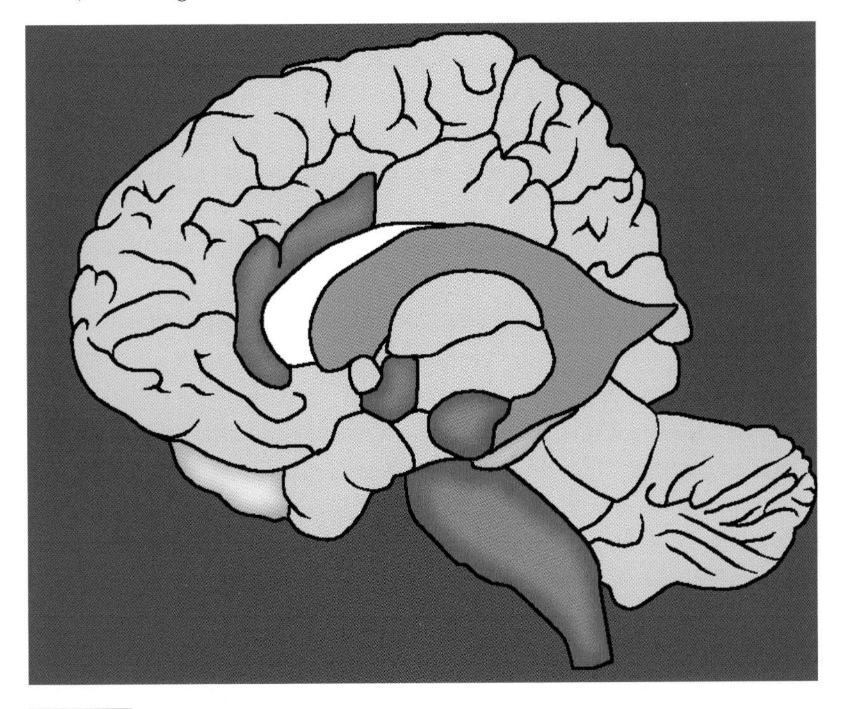

Figure 6.1 *Functional neuroanatomy of PTSD: in addition to activation of the amygdala and efferent circuitory, there is decreased activation of Broca's and perhaps other frontal areas, as well as impairment in hippocampus function (with apparent reduction in volume).*

in some work decreased hippocampus volume has correlated with trauma exposure or with cognitive impairment. Although it is possible that inherited variation in hippocampus volume may be a risk factor for subsequent PTSD (Lyons et al, 2001), more commonly such loss of volume is thought to represent atrophy (see below).

In healthy controls, imaging studies have demonstrated subcortical processing of masked emotional stimuli by the amygdala. Indeed, in an early study PTSD patients exposed to audiotaped traumatic and neutral scripts during PET were found to have increases in normalized blood flow in right-sided limbic, paralimbic, and visual areas, with decreases in left inferior frontal and middle temporal cortex (Rauch et al, 1996). Subsequent work has been more or less consistent (Rauch et al, 1998).

The authors of this research concluded that emotions associated with the PTSD symptomatic state are mediated by the limbic and paralimbic systems within the right hemisphere, with activation of visual cortex perhaps corresponding to visual re-experiencing. Decreased activity in Broca's area during exposure to trauma in PTSD, however, is consistent with patients' inability to process verbally traumatic memories (Rauch et al, 1996).

A recent study was suggestive of decreased benzodiazepine receptor binding in prefrontal cortex of PTSD patients (Bremner et al, 2000a). There is also a growing empirical literature on the anterior cingulate in particular in PTSD, with some data supporting an hypothesis of decreased activity in this region (Hamner et al, 1999). A recent study, for example, demonstrated that the ratio of N-acetylaspartate to creatinine, a marker of neural integrity, was significantly lower in the anterior cingulate of children and adolescents with PTSD than in healthy controls (De Bellis et al, 2000a). (Interestingly, although a conversion disorder, like PTSD, involves a shift from verbal to non-verbal processes, this condition appears to involve a somewhat different functional neuroanatomy. Halligan et al, 2000).

Although involvement of the basal ganglia has not typically been found in functional imaging studies of PTSD, a study of single-photon emission computed tomography (SPECT) scans in patients with PTSD and OCD reported that these groups had similarities in comparison to scans of patients with panic disorder and healthy controls (Lucey et al, 1997). The authors suggested that this might reflect the existence of repetitive intrusive symptoms in both PTSD and OCD. The possibility of certain phenomenological and psychopharmacological similarities between PTSD and OCD certainly bears further study. Also, interactions between the amygdala and corticostriatal systems may be important in mediating the transition from emotional reaction to emotional action (Le Doux, 1998).

Neurocircuitry and neurochemistry

Animal studies have demonstrated that serotonin is involved in regulation of the amygdala and connecting structures at a number of points (Coplan and Lydiard, 1998); this may well be relevant to the mediation of PTSD symptoms. Clinical studies of abnormal paroxetine binding in PTSD, and exacerbation of PTSD symptoms in response to administration of the serotonin agonist mCPP are also consistent with a role for serotonin in this disorder (Connor and Davidson, 1998).

Furthermore, there is increasing evidence for the efficacy of SSRIs in the treatment of PTSD (Stein et al, 2000), with some hints that these agents may even be more effective than other classes of medication (Penava et al, 1996) (although there are no 'head-to-head' trials to decide this question). To date there are few studies of the effects of SSRIs on the functional neuroanatomy of PTSD. Nevertheless, preliminary evidence indicates that they exert an effect by normalizing temperolimbic activation (Seedat et al, 2000) (Figure 6.2).

A range of neurochemical findings in PTSD are consistent with sensitization of various neurotransmitter systems (Charney et al, 1993). In particular, there is evidence of hyperactive noradrenergic function, as well as dopaminergic sensitization. Such sensitization is also consistent with the role of environmental traumas in PTSD; it turns out that dopamine agonists and environmental traumas act as cross-sensitizers of each other. There is evidence that the amygdala and related limbic regions may play a particularly important role in the final common pathway of such hyperactivation.

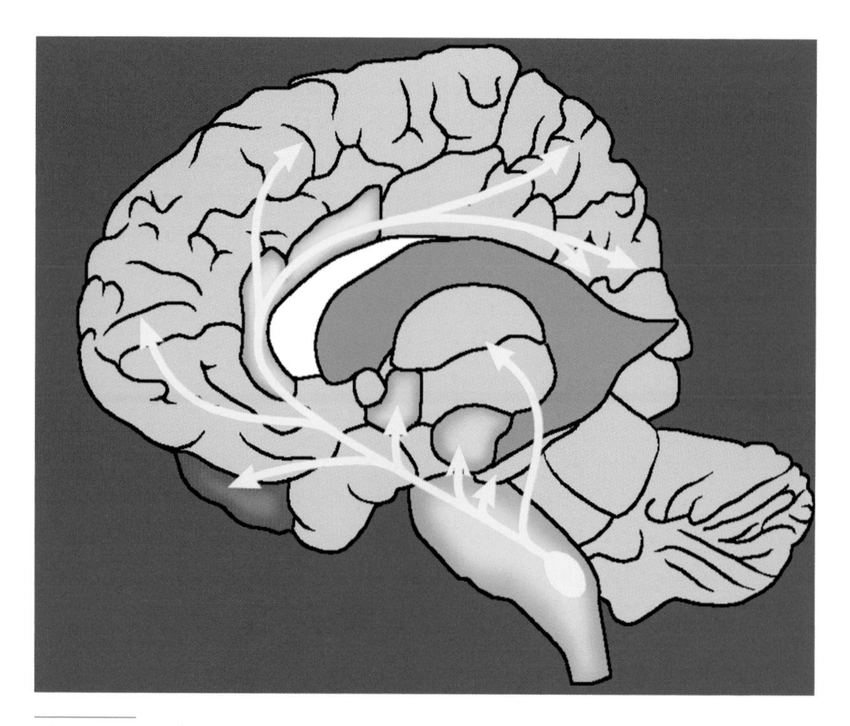

Figure 6.2 *Effects of SSRIs on the functional neuroanatomy of PTSD; normalization of activity in Broca's area and amygdala-mediated neurocircuits.*

Mild adrenergic activation may improve cognitive function (Cahill et al, 1994), with greater activation proving impairing (Steere et al, 1996). Interesting preliminary data indicate that administration of a β-blocker in the days following a traumatic event may help prevent the onset of PTSD (Pitman et al, 2002). PET scanning undertaken after administration of yohimbine demonstrated a significant increase in anxiety in PTSD patients, together with a decrease in several brain regions including prefrontal, orbitofrontal, temporal, and parietal cortex (Bremner et al, 1997a). This is perhaps consistent with some previous literature suggesting that during intense anxiety states there is a decrease in cerebral blood flow.

Little imaging work has been undertaken to date on the dopamine system in PTSD. However, in the prefrontal cortex of primates dopamine appears the most responsive system to stress; stress impairs prefrontal cognitive function, and this is ameliorated by pretreatment with low doses of dopamine blockers and other agents that reduce prefrontal dopamine turnover (clonidine, naloxone) (Arnsten and Goldman-Rakic, 1998). The authors of this study suggested that stress may take the prefrontal cortex 'off-line' to allow more habitual responses mediated by subcortical structures to regulate behavior.

Amygdala glutamate receptors and the N-methyl-D-aspartate receptor are likely to be involved in the neuronal mechanisms (e.g. long-term potentiation) that underlie fear conditioning, as well as in the extinction of fear-associated memories. To date no specific glutamatergic agent has reached the market for the treatment of PTSD or other anxiety disorders, although there is evidence that the anticonvulsant lamotrigine, which has glutamatergic effects, may be effective for the treatment of PTSD (Hertzberg et al, 1999). Future work in this direction seems potentially useful.

Another important set of neurochemical findings in PTSD has focused on the HPA system. PTSD is characterized by decreased plasma levels of cortisol, as well as increased glucocorticoid receptor responsiveness, suggesting that enhanced negative feedback may play a crucial role in the pathogenesis of the disorder (Yehuda et al, 1993). Such findings differ markedly from those found in chronic stress (where there is erosion of negative feedback and down-regulation of glucocorticoid receptors), in other anxiety disorders, and in depression. Notably, there are also high concentrations of cortisol-releasing factor receptors in amygdala, particularly in the central nucleus.

One important implication of the HPA findings in PTSD is the possibility that dysfunction in this system results in neuronal damage, particularly to the hippocampus. Animal studies have documented hippocampal damage after exposure to either glucocorticoids or naturalistic psychosocial stressors (Sapolsky, 2000). Parallel neurotoxicity in human disorders characterized by extreme stress could account for some of the cognitive impairments that are characteristic of this disorder.

Evolutionary considerations: going 'off-line'

The ability of the brain–mind to go 'off-line' in response to trauma presumably has ancient phylogenetic roots, having evolved as an adaptive response. Nevertheless, it seems that in PTSD this process continues to be maintained even once the danger has passed.

At a neurobiological level this may reflect sensitization of neurochemical systems, perhaps with consequent damage to the hippocampus. Essentially, excessive trauma has resulted in dysfunction. In this perspective, a crucial risk factor for PTSD would be previous exposure to traumatic events.

A psychobiological conceptualization may emphasize that in humans language and higher cognitive processes ordinarily play an important role. When these higher functions are taken 'off-line', there is an inability to process traumatic events. In PTSD, this problem persists.

Risk factors for PTSD (Table 6.2) can readily be conceptualized in terms of such a view. Thus patients with pre-trauma poor processing skills may be more prone to develop PTSD. Similarly, patients with peritraumatic dissociation are more likely to have difficulty in verbalizing their responses. Finally, patients who experience guilt, shame, or lack of social support in the aftermath of traumatic events may have more difficulty in processing such experiences (Yehuda, 1999).

Management

It is crucial to establish a safe and secure environment for the traumatized patient. While for some patients the danger is past, for others (such

Table 6.2 Risk Factors for PTSD

Pre-trauma: ↓IQ, ↑threat perception.
Peri-trauma: dissociation, negative view of the trauma
Post-trauma: anger, guilt, shame, ↓support

as those subject to domestic violence) threats may be ongoing. Similarly, the establishment of a trusting psychotherapeutic relationship is key; patients understandably often seem to require evidence that their clinician is reliable and dependable before they fully commit themselves to a treatment.

There is now strong support for the value of pharmacotherapy in PTSD. Early in the history of the treatment of this condition, medication was seen as a tool for encouraging anamnesis. The shift towards seeing PTSD as a medical disorder, characterized by specific psychobiological dysfunctions, has encouraged randomized controlled trials. While a number of different agents have been found effective, the most evidence comes from work on SSRIs, and these are currently viewed as the first-line agents of choice (Stein et al, 2000). Benzodiazepines, although commonly used, do not in fact appear helpful.

There is strong evidence for the value of cognitive-behavioral psychotherapy (CBT) in the management of PTSD. Interestingly, there is a good deal of overlap between the principles of CBT and those of psychodynamic psychotherapy in the management of traumatized patients. Both encourage exploration of the traumatic event, with a gradual reduction in avoidance behaviors. The re-telling of the trauma presumably allows an integration of implicit somatic and explicit verbal memories, and the articulation of new schemas that incorporate the traumatic experience into the patient's worldview.

As detailed earlier, preclinical data on fear conditioning provide a persuasive model for understanding the value of both pharmacotherapeutic and psychotherapeutic interventions. There is also a small but growing literature on the optimal use of both pharmacotherapy and psychotherapy in PTSD (Southwick and Yehuda, 1993). It is interesting to note, however, that in the acute aftermath of trauma, preliminary evidence for pharmacotherapeutic prophylaxis is promising, whereas psychotherapeutic intervention at this stage may be unhelpful (Pitman et al, 2002).

Many studies of PTSD have been undertaken in Vietnam veterans, a group that sometimes demonstrated significant treatment-refractoriness. Whether such data can be extrapolated to other groups of veterans or civilians is debatable. Nevertheless, a significant portion of PTSD patients does not respond to first-line pharmacotherapy or psychotherapy. There is a small literature on the use of augmentation and combination strategies for treatment-refractory PTSD (Stein et al, 2000); more research in this area is needed.

Conclusion

Some authors have argued that 'trauma' ultimately constitutes the final common pathway that lies at the bottom of all psychopathology. Freud, of course, reversed his early similar stance, and modern data on individual psychobiological susceptibilities suggest that these also play a crucial role in determining whether responses to trauma are characterized by resilience and growth, or by psychopathology and dysfunction.

Nevertheless, the fact that traumatic processing can occur both implicitly and explicitly may point to a number of significant truths about human nature. Such a division reflects contrasts between mind (explicit cortical articulation) and body (implicit limbic activation), and between head (explicit) and heart (implicit). Through verbalization (i.e. frontal processing), there is an integration, and a healing. Such a view would certainly seem to be consistent with a range of psychodynamic thinking (Horowitz, 1991).

Indeed, for cognitive-affective models of the mind, trauma may be a crucial phenomenon. Trauma determines whether human processing becomes 'hot' and 'irrational' (Greenberg and Safran, 1990), whereas computer processing always remains 'cool' and 'logical'. The advantage of 'hot processing' is perhaps that humans can work through their traumas, becoming resilient and creative in ways that computers cannot. The disadvantage is perhaps that humans run the risk of responding to trauma with dissociation and psychopathology.

chapter 7

Social anxiety disorder

Symptoms and assessment

Social anxiety disorder (SAD; social phobia or SP) is, apart from specific phobia, the most common of the anxiety disorders, with prevalence ranging from 3 to 16% in various studies (Davidson et al, 1993b; Kessler et al, 1994). It is more common in women in community studies, but in clinical studies the proportion of men with the disorder increases considerably. Cross-national community studies show similarities in demographic and clinical features in different parts of the world (Weissman et al, 1996).

SAD is characterized by fear of embarrassment or humiliation in social situations (Table 7.1). Social situations comprise social interaction (e.g. talking in small groups, dating) and performance (e.g. speaking or eating in front of others). These situations are associated with symptoms of panic, but the panic attacks of SAD are more likely to be characterized by blushing, tremor, and averted gaze (Amies et al, 1983).

As in other anxiety disorders, the anxiety often leads to avoidance and disability. People with SAD are more likely to remain unmarried, to drop out of school or college and to earn less money than those without these conditions. The term 'social anxiety disorder' is increasingly recommended as possibly less stigmatizing than 'social phobia' and helps emphasize the overlap between different anxiety disorders.

Important subtypes of SAD are generalized and discrete SAD. General-

Table 7.1 Symptoms of SAD (modified from DSM-IV-TR).

(A) A marked and persistent fear of one or more social or performance situations in which the person is exposed to unfamiliar people or to possible scrutiny by others. The individual fears that he or she will act in a way (or show anxiety symptoms) that will be humiliating or embarrassing.
(B) Exposure to the feared social situation almost invariably provokes anxiety, which may take the form of a situationally bound or situationally predisposed panic attack.
(D) The feared social or performance situations are avoided or else are endured with intense anxiety or distress.
(E) The avoidance, anxious anticipation, or distress in the feared social or performance situation(s) interfere significantly with the person's normal routine, occupational (academic) functioning, or social activities or relationships, or there is marked distress about having the phobia.
(F) The fear or avoidance is not due to the direct physiological effects of a substance (e.g. a drug of abuse, a medication) or a general medical condition and is not better accounted for by another mental disorder.

Specify if:
Generalized: if the fears include most social situations.

ized SAD is characterized by fear of most social situations, whereas discrete SAD is limited to one or two performance situations. Generalized SAD is associated with more severe and disabling symptoms, and may be more familial. It is possible, however, that there exists a continuum from generalized SAD to discrete SAD to fear of public speaking (Kessler et al, 1998).

It is important to note the comorbidity of SAD with depression and with substance use disorders (Kessler et al, 1999). Indeed, given that SAD often begins early in life, has a chronic course, and usually precedes other disorders, it is fair to argue that SAD predisposes to such conditions. Interestingly, the symptoms of depression in SAD are often atypical (characterized by hyperphagia, hypersomnia, leaden paralysis, and rejection sensitivity).

Symptoms of SAD can be measured using a number of different scales, including the Liebowitz Social Anxiety Scale (Liebowitz, 1987) (Table A.6, see Appendix). This scale measures the extent of both fear and avoidance for a range of different social and performance situations.

Cognitive-affective considerations: social cognition

Social anxiety is of course a normal and adaptive emotion. Indeed, this is arguably the reason it has taken so long to recognize SAD as a psychiatric disorder, and why it continues to remain underdiagnosed and undertreated (Schneier et al, 1992). Clinicians, familiar of course with their own feelings of social anxiety, may mistakenly 'normalize' their patients' symptoms of SAD.

Indeed, although textbooks of psychopathology often quote the example of the lion and the antelope to illustrate human fright–fight–flight responses, humans are very social primates and threatening conspecifics (i.e. other humans) are perhaps more constantly relevant than is the occasional predator. Humans have presumably evolved efficient mechanisms for recognizing thoughts and emotions in conspecifics and for responding accordingly.

There is a growing body of knowledge about the neurocircuitry involved in recognizing and processing the faces, emotions, and gaze of others (Allison et al, 2000). A range of structures have been suggested to mediate social cognition, including the amygdala and temporal regions, the striatum, and prefrontal and cingulate cortex (Adolphs, 2001).

There is also a growing body of knowledge about the neuroanatomy of anxiety in general, as reviewed earlier. Nevertheless, relatively little is known about the neuroanatomy of social anxiety per se. Humans, as Darwin pointed out (Darwin, 1965), are the only animals that blush, so that while the general neuroanatomy of anxiety may be relevant, other more specific circuits also require delineation.

Amygdala–dorsal striatum ganglia circuitry, for example, may mediate inhibitory avoidance and motor learning (Davis and Whalen, 2001);

interestingly, socially anxious children show reduced general facial activity and have a more restricted facial repertoire (Melfsen et al, 2000). Also, amygdala–ventral striatal circuitry may be relevant in so far as it bridges processing of affects and appetitive behavior (Davis and Whalen, 2001)

Shyness and behavioral inhibition (manifesting in social situations) may represent predisposing traits for SAD, and are thought to have a heritable component. Although the neurobiology of behavioral inhibition is incompletely understood, it might be speculated that prefrontal hyperactivity may contribute (Johnson et al, 1999; Kagan et al, 1988).

Neurocircuitry of SAD

Given that social anxiety is a form of anxiety, it would not be surprising if the amygdala played a role in mediating SAD (Figure 7.1). Certainly, a

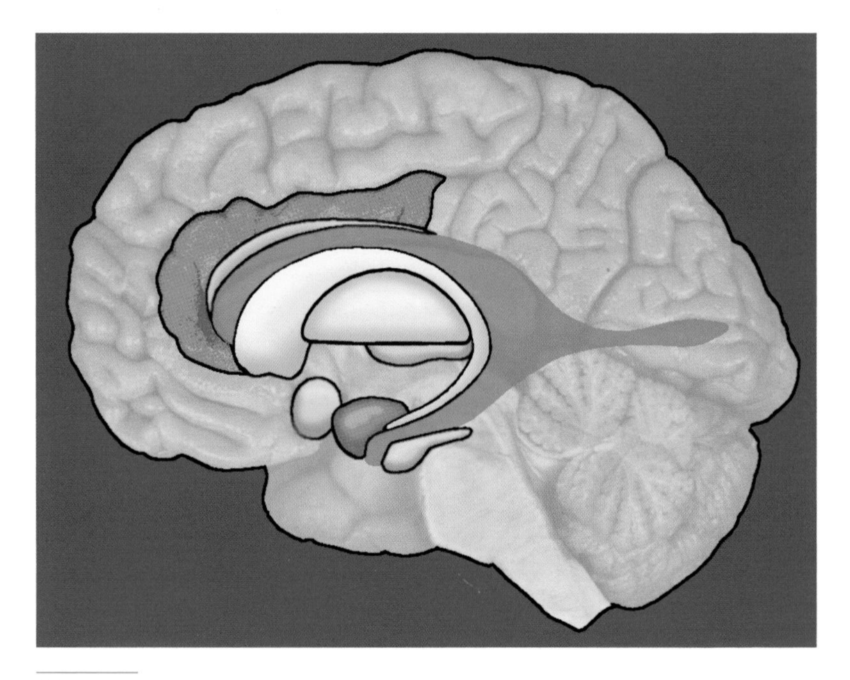

Figure 7.1 *Functional neuroanatomy of SAD: increased amygdala and cingulate activity, with decreased basal ganglia activity.*

range of panicogenics are able to trigger panic attacks in SAD patients, albeit not to the same extent as in panic disorder (Stein et al, 2002). Furthermore, there is interesting recent data showing that SAD patients demonstrate selective activation of the amygdala when exposed to fear-relevant stimuli (Birbaumer et al, 1998) or tasks (Tillfors et al, 2001), or show abnormal patterns of amygdala activation during aversive conditioning (Schneider et al, 1999). Conversely, after lesions of the amygdala (Klüver–Bucy syndrome) there may be inappropriate loss of social fear.

There is also growing evidence that striatal neurocircuits play a role in SAD. Thus, patients with SAD have a greater reduction in putamen volume with aging (Potts et al, 1994), reduced choline and creatinine signal-to-noise ratios in subcortical, thalamic, and caudate areas (Davidson et al, 1993a), and decreased N-acetyl-aspartate (NAA) levels and a lower ratio of NAA to other metabolites in cortical and subcortical regions (Davidson et al, 1993a; Tupler et al, 1997). Furthermore, striatal dopamine systems may be abnormal in SAD (see below).

Finally, frontal areas may play a role in social anxiety. Although not all work is consistent, there is a report of increased dorso-lateral prefrontal cortex activity during symptom provocation in a PET study of SAD (Nutt et al, 1998), and of cortical gray matter abnormalities in SAD in particular (Tupler et al, 1997). Anterior cingulate, which is involved in performance monitoring (McDonald et al, 2000), may play a crucial role in a number of anxiety disorders, including SAD. In addition, imaging studies that have pooled or compared findings across different anxiety disorders suggest the importance of increased activation of inferior cortex in mediating anxiety symptoms (Rauch et al, 1997b).

Neurocircuitry and neurochemistry

Given the potential involvement of these various regions (amygdala, basal ganglia, frontal) in SAD, it can be postulated that the serotonin system plays an important role in the mediation of social anxiety. As reviewed in Chapters 4 and 5, the serotonin system branches widely and extends to both amygdala and cortico-striatal neurocircuits (Figure 1.1).

Furthermore, serotonin plays a central role in mediating social behavior in animal models. Thus, reduction of serotonin function leads to avoidance of affiliative social behaviors in primates, while enhancement of serotonergic function results in increased pro-social behaviors (Raleigh et al, 1983). Similarly, free-ranging primates with low CSF 5-HIAA have less social competence and emigrate from their social groups earlier (Mehlman et al, 1995; Raleigh et al, 1985). Importantly, however, changes in status result in a change in serotonergic function–removal of dominant animals from the group results in significant decreases in serotonin levels (Raleigh et al, 1984), so that the relationships between social behavior and serotonin status are complex. Interestingly, there are some data to suggest that increasing serotonergic activity in humans is also associated with an increase in social affiliation, although the data are not altogether consistent.

There is little evidence for abnormalities in static peripheral measures of serotonin function in SAD (Stein et al, 1995; Tancer et al, 1994). Nevertheless, early pharmacological 'challenge' studies with serotonergic agents, which assess the dynamic responsiveness of the serotonergic system, provided some support for serotonin dysfunction in SAD (Tancer et al, 1994). A low-activity polymorphism of the serotonin transporter (5-HTTP) gene may be associated with anxiety-related traits, although this particular polymorphism does not appear relevant to SAD (Stein et al, 1998).

SSRIs are increasingly seen as the medication treatment of choice for SAD (Ballenger et al, 1998; van der Linden et al, 2000). Relatively little is known about the effects of SSRIs on the functional neuroanatomy of SAD. Nevertheless, treatment with SSRIs may be hypothesized to normalize dysfunctional circuitry in SAD. Certainly, there is some evidence that SSRI treatment leads to reduced activity in amygdala-hippocampal, frontal, and cingulate regions in SAD patients (van der Linden et al, 1999; Furmark et al, 2002) (Figure 7.2).

A range of evidence points to the involvement of the dopaminergic system in SAD. Animal work, for example, shows that dopamine levels are reduced in timid mice (Mayleben et al, 1992) and that striatal D_2 binding is decreased in lower social status monkeys (Grant et al, 1998).

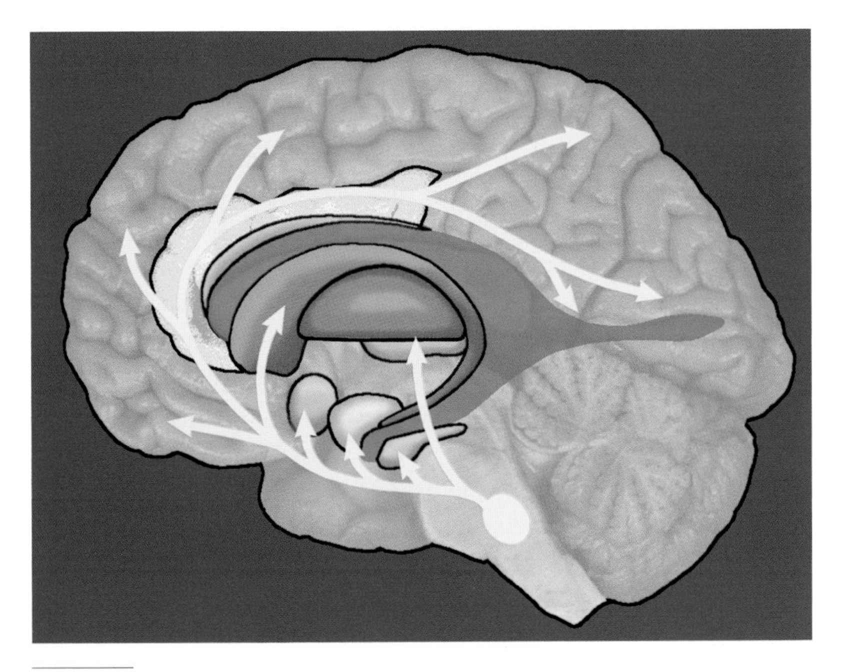

Figure 7.2 *Effect of SSRIs on functional neuroanatomy of SAD: normalized activity in amygdala, cingulate and basal ganglia.*

As in the case of serotonin, the relationships between social status and dopamine function are complex (rather than simply unidirectional).

In depressed patients, low CSF dopamine correlated with a measure of introversion and CSF homovanillic acid (HVA) was lower in panic disorder patients with SAD than in those without. Furthermore, there may be increased SAD in patients who subsequently develop Parkinson's disease (Richard et al, 1996). Conversely, patients treated with dopamine blocking agents may develop an increase in social anxiety symptoms (Pallanti et al, 1999). Indeed, monoamine oxidase inhibitors are effective in the treatment of SAD and rejection sensitivity in atypical depression, whereas tricyclic antidepressants are not useful for these indications.

There is an association between decreased dopamine D_2 receptor agonist reactivity and reduced 'positive emotionality', between low striatal D_2 binding or dopamine transport binding and detachment, between certain dopamine D_2 and dopamine transporter polymorphisms and schizoid/avoidant behavior, and perhaps between short dopamine D_4 alleles and decreased novelty seeking (although findings are not entirely consistent, and it is not yet clear to what extent such constructs relate to social anxiety per se) (Stein et al, 2002).

Functional brain imaging has provided the strongest evidence to date that the dopamine system plays a significant role in mediating SAD. A study of the density of dopamine reuptake sites found that striatal dopamine reuptake site densities were markedly lower in SAD patients than in normal controls (Tiihonen et al, 1997a). Furthermore, striatal D_2 receptor binding was lower in SAD than in controls (Schneier et al, 2001). Taken together, these findings would suggest that SAD is characterized by decreased dopamine function.

Other neurobiological systems that may be involved in SAD include certain neuropeptides (Insel, 1997), the HPA axis (Kagan et al, 1988), and perhaps growth hormone (Uhde, 1994). Nevertheless, such work has not yet led to any pharmacological interventions in the clinic.

Evolutionary considerations: appeasement alarms

Although it was Darwin who noted that humans are the only animals to blush, it was Twain who made the astute observation that only humans needed to (Twain, 1897). What indeed is the function of blushing? Given the close relationship between blushing and SAD, an answer to this question might shed light on the distal (evolutionary) mechanisms that underlie this condition.

In the animal world, dominant and submissive status are signaled by a range of mechanisms. Appeasement displays, for example, play an important role in indicating acceptance of the status quo to a dominant conspecific (De Waal, 1989). Is it possible that blushing serves as an appeasement display? Certainly, an embarrassed blush accompanied by lowering of the gaze and a silly grin does seem reminiscent of certain appeasement displays. Furthermore, empirical studies show that displays of embarrassment do mitigate the negative reactions of others (Leary et al, 1992).

It turns out that blushing and SAD have somewhat similar demographics (more common in females and in younger people) and that both are elicited by similar triggers (social attention) (Stein and Bouwer, 1997b). In addition, a range of data indicate that patients with SAD misperceive information about the need for social appeasement (e.g. exaggerated view of the low status of the self, overestimation of social threat). Although the neurobiology of blushing remains incompletely understood, it is possible to speculate that there are certain overlaps with that of SAD.

Evolutionary hypotheses about false alarms are strengthened when cases of the pathological absence of a particular alarm are noted. Are there some conditions in which there is insufficient social anxiety? Apart from Klüver–Bucy syndrome (in which there is a loss of fear), it turns out that people with a hereditary condition know as William's disorder may be characterized by hypersociability (Bellugi et al, 1999). Such hypersociability can of course potentially land people in all sorts of trouble! The neurobiology of this condition remains incompletely understood.

Management

The treatment of SAD begins, once again, with a psychoeducational model. Many patients view their symptoms as evidence of immutable personality traits. A different model, in which a range of different factors contribute to SAD, and according to which a number of different interventions are useful, needs to be encouraged. Such a model should be taught not only to patients, but also to primary practitioners, as SAD remains underdiagnosed and undertreated, despite being associated with increased medical utilization.

While the MAOIs have long been known effective in the treatment of SAD, these agents have a number of significant disadvantages. RIMAs are significantly better tolerated, although the literature has suggested that their effect size may be comparatively low. Similarly, although benzodiazepines may have a role in some patient, they are associated with important problems. Thus, the SSRIs are increasingly viewed as a first-line pharmacotherapy of choice in SAD. Interestingly, SSRIs may be useful not only for more generalized SAD, but also for less generalized SAD (Stein et al, 2001).

CBT has also been demonstrated effective for the treatment of SAD. Patients are encouraged to expose themselves to anxiety-inducing situations, and to gradually reduce their avoidance behaviors. Avoidance behaviors can be fairly subtle, for example, looking at the ground, or holding on to a lecturn when giving a talk. Over time, patients can learn to overcome their social anxiety, with consequent improvements in social and occupational dysfunction.

The recent finding that both pharmacotherapy and psychotherapy are able to normalize the functional neuroanatomy of SAD (Furmark et al, 2002) supports an integrative approach to the management of this disorder. It seems reasonable to consider pharmacotherapy in patients who are unable to tolerate the anxiety caused by exposure, and it seems reasonable to follow CBT principles even when treatment is initiated with a medication.

There is very little literature on the management of treatment refractory SAD. Reassessment for substance use and general medical disorders is always a useful step in the management of treatment-refractoriness. Interesting uncontrolled data suggest that venlafaxine may be useful in SAD patients who have failed to respond to SSRIs (Altamura et al, 1999). The classical MAOIs can also be considered in SAD patients who fail to respond to new generation antidepressants. There is a need, however, for further investigation in this area.

Conclusion

SAD is in some ways the most 'human' of the anxiety disorders. If panic and OCD have evolutionary origins in suffocation and grooming alarms then we might expect that other animals may also suffer from analogous symptoms. Arguably only higher primates, with their complex social organizations, would suffer from SAD. Indeed, it is this very 'humanness' that has perhaps contributed to the underdiagnosis and undertreatment of SAD.

A perennial question in cognitive science, with its reliance on computational models, is 'can a computer do x?'. This question is related to the classic Turing test (Turing, 1950), arguably first passed by PARRY, a computer program that modeled paranoid speech so well that interviewers did not know whether they were interacting with a computer or with a human (Colby, 1975). Modern computers are increasingly adept at cognitive tasks, including ordinary conversation. But can a computer feel 'socially anxious'?

The view taken in this volume would be that the fact of social anxiety points to important truths about the human brain–mind and the evolution of its mediating wetware. Given its complexity, it is perhaps unsurprising that sometimes social anxiety goes awry. As in the case of other psychiatric disorders, such dysfunction will rarely be the result of a single neurobiological disturbance, nor will it simply be a meaningful response to unusual circumstances.

When social anxiety goes awry, it may be argued that this dysfunction will be grounded not only in various neuroanatomical structures (including amygdala, striatum), but will also be rooted in an altered experience of social relations. From a clinical perspective, it is crucial to recognize the distress and impairment of those with SAD, and to provide appropriate intervention.

chapter 8

Conclusion

A first focus of this volume was on providing user-friendly updates on the neurocircuitry of depression and anxiety disorders, partly with the aim of providing a clinical foundation for approaching the symptoms and treatment of these conditions.

Both animal and clinical studies indicate that the amygdala and paralimbic structures play an important role in conditioned fear and in anxiety disorders. Amygdala lesions are classically associated with decreased fear response, and conversely, hyperactivation of the limbic system is characteristic of a number of different anxiety disorders. Paralimbic regions such as anterior cingulate appear to play a key role at the interface of cognition and emotion. The apparent centrality of such systems to different anxiety disorders may account in part for their high comorbidity. Additional features of limbic involvement may, however, be somewhat specific to particular disorders, for example, hippocampal shrinkage in PTSD, or parahippocampal asymmetry in panic disorder. Serotonin reuptake inhibitors are increasingly viewed as first-line treatments for anxiety disorders, and innervation of amygdala and paralimbic structures by serotonergic neurons may be crucial in explaining their efficacy. Future research is needed to delineate the underlying neurocircuitry of anxiety disorders in order to help treat those patients who currently remain refractory to first-line agents.

In addition, CSTC pathways may be important in both mood and anxiety disorders, particularly in OCD and certain putative obsessive-compulsive spectrum disorders, such as Tourette's syndrome. There is a growing consolidation of imaging, immunological, genetic, and treatment data around this model. It is particularly remarkable that CSTC pathways in OCD as well as in depression can be normalized by pharmacotherapy, psychotherapy, and by neurosurgery. In some ways, it can be argued that whereas OCD was once viewed as the key to a psychodynamic understanding of the mind, OCD and some obsessive-compulsive spectrum disorders such as Tourette's syndrome are now the neuropsychiatric disorders par excellence. Fortunately, the serotonin reuptake inhibitors, as well as a range of other medications, are able to bring relief to many patients with OCD and depression. Future research is needed in order to develop agents, perhaps working at a second messenger level or beyond (Lesch, 2001), that target symptoms more quickly and that provide novel options for refractory patients.

Owing to space limitations, this volume has paid little attention to issues of laterality. The discussion of depression noted, however, that left-sided lesions after stroke have been associated with depression while right-sided lesions have been associated with mania. Similarly, low positive affect has been associated with hypoactivation of left prefrontal cortex (Mineka et al, 1998). In panic disorder secondary to temporal lesions the right hemisphere is typically involved, and in PTSD right limbic activation is prominent. In OCD there is also some evidence of lateralization effects, with neurosurgical treatment more successful when lesions are made in the right hemisphere. The significance of such findings is currently somewhat unclear. Nevertheless, marked brain laterality has been suggested by a number of authors to be a defining feature of homo sapiens, and future work on the neurocircuitry of mood and anxiety disorders would do well to pay careful attention to this issue.

A second theme of this volume was to include not only proximate but also distal (evolutionary) psychobiology. Such a framework is not only theoretically important given that evolutionary theory is the ultimate basis for biology and neuroscience, but again may have clinical value.

The account here was based in the notion of 'false alarms' (Stein and Bouwer, 1997a). While depression and anxiety may ordinarily be a normal response, in psychiatric disorders they may represent a specific defect; nevertheless, the line between adaptive defence and maladaptive defect may not always be a sharp one. In particular, it has been hypothesized that evolutionary-based neurobiologically grounded alarm systems are erroneously triggered in these conditions. The argument here assumes that different kinds of alarm systems have evolved in order to respond to different kinds of dangers; such an assumption then leads to hypotheses linking particular clinical disorders with specific dangers (e.g. panic disorder and suffocation). This view allows a focus on both environmental experience and neurobiological underpinnings in considering both pathogenesis and intervention.

Aspects of the evolutionary approach remain, however, speculative. It is also pertinent to acknowledge that human evolution has led to a unique semiotic capacity in our species, and that a cognitive level of analysis is therefore central to a comprehensive understanding of ordinary and pathological behavior (Stein and Bouwer, 1997a). On the one hand, evolutionary theory may entail a rethinking of many psychosocial phenomena including the way in which some anthropologists have seen anxiety symptoms in different cultures as radically different (Stein and Williams, 2002). On the other hand, a cognitive level of analysis cannot be neglected if human behavior in general, and the anxiety disorders in particular, are to be understood fully.

Some evidence does seem to be accumulating for the various 'false alarm' hypotheses presented in this volume; the argument that panic disorder represents a false suffocation alarm, for example, now has a range of empirical support. Further work is, however, needed in order to consolidate an evolutionary approach to the mood and anxiety disorders. In the interim, this approach may be useful in the clinical situation – in allowing the clinician to understand symptoms more empathically and in providing a meaningful explanatory model of the disorder that may help the patient make sense of his or her experience.

A third focus of this volume was on providing models that allow an integrative approach to conceptualizing the brain–mind and its psychopathology. Depression and anxiety are perhaps particularly useful phenomena to consider in so far as they seem to demand such an integrative approach.

Certainly, depression and anxiety are significant and complex emotional phenomena which cannot be accounted for by the outdated meta-psychology of psychoanalysis, by simplistic cognitivist accounts that specify the brain–mind solely using computational metaphors, or by reductionistic attempts to account for mental phenomena in terms of biological markers (see Chapter 1). Rather, depression and anxiety can be conceptualized as impassioned forms of cognitive processing; they are grounded in the wetware of the brain–mind and they develop and are played out within the context of social interactions. Indeed, cognitive-affective neuroscience will ultimately need to delineate these emotions and their disorders in even greater detail if the brain–mind is to be understood better.

Playwrights and novelists have long drawn a distinction between emotion and reason, heart and mind. Freud was a pioneer in so far as he attempted to bring a scientific account to the passions, emphasizing the importance of unconscious processing. He contrasted the compulsive disorders (where affect was repressed) and the hysterical disorders (where thought was repressed). His contribution was to describe in great detail how different kinds of psychopathologies reflected different configurations of passion and reason (Table 8.1). While this work was invaluable, Freud's model of the mind was of course based on nineteenth-century physics, and the metaphor of energy being expressed, repressed, or redirected could only go so far (Stein, 1992). Early on, however, Janet, Piaget, and others demonstrated that psychodynamic constructs could also be reframed in more cognitive-affective terms; rather than discarding early insights, these should arguably be specifically included in modern texts on cognitive-affective neuroscience to spur interest in complex brain–mind phenomena (Stein, 1997).

Table 8.1 Freudian perspective on obsessional and hysterical neurosis.

Hysterical neurosis	*Obsessional neurosis*
Repression of ideas	Repression of affect

Indeed, today we are closer than ever to having the kind of (proximate and distal) psychobiological understanding of cognition and affect of which Freud once dreamed. In this volume, we have suggested, for example, that OCD is characterized by a form unconscious processing in which there is conscious intrusion of procedural strategies, whereas PTSD is characterized by a form of non-cognitive processing in which implicit and explicit processes remain dissociated (Table 8.2). Of course, much that Freud and his followers taught retains its validity; the idea that impulsivity underlies compulsivity in OCD remains a possibility, and the modern emphasis on processing of trauma remains at its core a psychodynamic idea. Nevertheless, there have also been significant advances; among the most important are the development of an empirical research program that puts theoretical constructs and clinical interventions to the test. The future of cognitive-affective neuroscience research seems bright, and we can therefore expect continued advances in clinical practice in general, and in the mood and anxiety disorders in particular.

Table 8.2 A cognitive-affective neuroscience perspective on OCD and PTSD.

PTSD	*OCD*
Dysfunction in fear conditioning	Dysfunction in procedural strategies
Amygdala-mediated mechanisms	CSTC-mediated mechanisms
Treatment involves trauma processing	Treatment involves exposure to stimuli

References

Adolphs R. The neurobiology of social cognition. *Curr Opin Neurobiol* 2001;**11**:231–239

Alexander CE, DeLong MR, Strick PL. Parallel organization of functionally segregated circuits linking basal ganglia and cortex. *Ann Rev Neurosci* 1986;**9**:357–381

Allison T, Puce A, McCarthy G. Social perception from visual cues: role of the STS region. *Trends Cognitive Sci* 2000;**4**:267–277

Altamura AC, Pioli R, Vitto M, et al. Venlafaxine in social phobia: a study in selective serotonin reuptake non-responders. *Int Clin Psychopharmacol* 1999; **14**:239–245

American Psychiatric Association. *Diagnostic and Statistical Manual of Mental Disorders*, 3rd Edn. Washington, DC: American Psychiatric Press, 1980

Amies PL, Gelder MG, Shaw PM. Social phobia: a comparative clinical study. *Br J Psychiatry* 1983;**142**:174–179

Arnsten AFT, Goldman-Rakic PS. Noise stress impairs prefrontal cortical cognitive function in monkeys: evidence for a hyperdopaminergic mechanism. *Arch Gen Psychiatry* 1998;**55**:362–368

Austin M-P, Mitchell P, Goodwin GM. Cognitive deficits in depression: possible implications for functional neuropathology. *Br J Psychiatry* 2001;**178**:200–206

Ballenger JC, Davidson JA, Lecrubier Y, et al. Consensus statement on social anxiety disorder from the international consensus group on depression and anxiety. *J Clin Psychiatry* 1998;**59**:54–60

Ballenger JC, Davidson JR, Lecrubier Y, et al. Consensus statement on generalized anxiety disorder from the International Consensus Group on Depression and Anxiety. *J Clin Psychiatry* 2001;**62S**:53–58

Bandelow B. The use of the Panic and Agoraphobia Scale in a clinical trial. *Psychiatry Res* 1998;**77**:43–49

Barr CL, Goodman WK, Price LH, et al. The serotonin hypothesis of obsessive compulsive disorder: implications of pharmacologic challenge studies. *J Clin Psychiatry* 1992;**53S**:17–28

Baumann B, Bielau H, Kreil D, et al. Circumscribed numerical deficit of dorsal raphe neurons in mood disorders. *Br J Psychiatry* 2002;**32**:93–104

Baxter LR, Schwartz JM, Bergman KS, et al. Caudate glucose metabolic rate changes with both drug and behavior therapy for OCD. *Arch Gen Psychiatry* 1992;**49**:681–689

Bechara A, Tranel D, Damasio H, et al. Double dissociation of conditioning and declarative knowledge relative to the amygdala and hippocampus in humans. *Science* 1995;**269**:1115–1118

Beck AT. *Depression: Clinical, Experimental, and Theoretical Aspects*. New York: Harper & Row, 1967

Becker G, Berg D, Lesch KP, et al. Basal limbic system alteration in major depression: a hypothesis supported by transcranial sonography and MRI findings. *Int J Neuropsychopharmacol* 2001;**4**:21–31

Bell C, Forshall S, Adrover M, et al. Does 5-HT restrain panic? A tryptophan depletion study in panic disorder patients recovered on paroxetine. *J Psychopharmacol* 2002;**16**:5–14

Bell C, Nutt DJ. Tryptophan depletion and its implications for psychiatry. *Br J Psychiatry* 2001;**178**:399–405

Bellugi U, Adolphs R, Cassady C, et al. Towards the neural basis for hypersociability in a genetic syndrome. *Neuroreport* 1999;**10**:1653–1657

Bench CJ, Friston KJ, Brown RG, et al. Regional cerebral blood flow in depression measured by positron emission tomography: the relationship with clinical dimensions. *Psychol Med* 1993;**23**:579–590

Bhaskar R. *A Realist Theory of Science*, 2nd Edn. Sussex: Harvester Press, 1978

Birbaumer N, Grodd W, Diedrich O, et al. fMRI reveals amygdala activation to human faces in social phobics. *Neuroreport* 1998;**9**:1223–1226

Blumberg HP, Stern E, Martinez D, et al. Increased anterior cingulate and caudate activity in bipolar mania. *Biol Psychiatry* 2000;**48**:1045–1052

Bonne O, Brandes D, Gilboa A, et al. Longitudinal MRI study of hippocampal volume in trauma survivors with PTSD. *Am J Psychiatry* 2001;**158**:1248–1251

Bowlby J. *Attachment and Loss*, Vol 3. New York: Basic Books, 1980

Boyer W. Serotonin uptake inhibitors are superior to imipramine and alprazolam in alleviating panic attacks: a meta-analysis. *Int Clin Psychopharmacol* 1995;**10**:45–49

Bremner JD, Innis RB, Ng CK, et al. Positron emission tomography measurement of cerebral metabolic correlates of yohimbine administration in combat-related posttraumatic stress disorder. *Arch Gen Psychiatry* 1997a;**54**:246–254

Bremner JD, Innis RB, Salomon RM, et al. Positron emission tomography measurement of cerebral metabolic correlates of tryptophan depletion-induced depressive relapse. *Arch Gen Psychiatry* 1997b;**54**:364–374

Bremner JD, Innis RB, Southwick SM, et al. Decreased benzodiazepine receptor binding in prefrontal cortex in combat-related posttraumatic stress disorder. *Am J Psychiatry* 2000a;**157**:1120–1126

Bremner JD, Innis RB, White T, et al. SPECT [I-123]iomazenil measurement of the benzodiazepine receptor in panic disorder. *Biol Psychiatry* 2000b;**47**:96–106

Brewin CR. A cognitive neuroscience account of posttraumatic stress disorder and its treatment. *Behav Res Ther* 2001;**39**:373–393

Brody AL, Saxena S, Stoessel P, et al. Regional brain metabolic changes in patients with major depression treated with either paroxetine or interpersonal therapy. *Arch Gen Psychiatry* 2001;**58**:631–640

Buchsbaum MS, Hazlett E, Sicotte N, et al. Topographic EEG changes with benzodiazepine administration in generalized anxiety disorder. *Biol Psychiatry* 1985;**20**:832–842

Burgess N, Jeffrey KJ, O'Keefe J. *The Hippocampal and Parietal Foundations of Spatial Cognition*. New York: Oxford University Press, 1999

Byrum CE, Ahearn EP, Krishnan KR. A neuroanatomic model for depression. *Prog Neuropsychopharmacol Biol Psychiatry* 1999;**23**:175–193

Cahill L, Prins B, Weber M, et al. Beta-adrenergic activation and memory for emotional events. *Nature* 1994;**371**:702–704

Cendes FAFGP, Gambardella A, Lopes-Cendes I, et al. Relationship between atrophy of the amygdala and ictal fear in temporal lobe epilepsy. *Brain* 1994;**117**:739–746

Charney DS. Monoamine dysfunction and the pathophysiology and treatment of depression. *J Clin Psychiatry* 1998;**59S**:11–14

Charney DS, Deutch AY, Krystal JH, et al. Psychobiologic mechanisms of posttraumatic stress disorder. *Arch Gen Psychiatry* 1993;**50**:295–305

Charney DS, Heninger GR, Breier A. Noradrenergic function in panic anxiety: effects of yohimbine in healthy subjects and patients with agoraphobia and panic disorder. *Arch Gen Psychiatry* 1984;**41**:751–763

Cheyette SR, Cummings JL. Encephalitis lethargica: lessons for contemporary neuropsychiatry. *J Neuropsych Clin Neurosci* 1995;**7**:125–135

Colby KM. *Artificial Paranoia: A Computer Simulation of Paranoid Processes*. New York: Pergamon, 1975

Connor KM, Davidson JR. Further psychometric assessment of the TOP-8: a brief interview-based measure of PTSD. *Depress Anxiety* 1999;**9**:135–137

Connor KM, Davidson JRT. The role of serotonin in posttraumatic stress disorder: Neurobiology and pharmacotherapy. *CNS Spectrums* 1998;**3**:43–51

Coplan JD, Gorman JM, Klein DF. Serotonin related functions in panic-anxiety: a critical overview. *Neuropsychopharmacol* 1992;**6**:189–200

Coplan JD, Lydiard RB. Brain circuits in panic disorder. *Biol Psychiatry* 1998;**44**:1264–1276

Damasio AR. The somatic marker hypothesis and the possible functions of the prefrontal cortex. *Philos Trans R Soc* 1996;**351**:1413–1420

Darwin C. *The Expression of Emotion in Man and Animals*. Chicago, IL: Chicago University Press, 1965:1872

Davidson JRT, Boyko O, Charles HC, et al. Magnetic resonance spectroscopy in social phobia. *J Clin Psychiatry* 1993a;**54S**:19–25

Davidson JRT, Hughes DL, George LK, et al. The epidemiology of social phobia: findings from the Duke Epidemiologic Catchment Area study. *Psychological Med* 1993b;**23**:709–718

Davidson RJ. Asymmetric brain function, affective style and psychopathology: the role of early experience and plasticity. *Devel Psychopathol* 1994;**6**:741–758

Davidson RJ, Putnam KM, Larson CL. Dysfunction in the neural circuitry of emotion regulation – a possible prelude to violence. *Science* 2001;**289**:591–594

Davis M, Whalen PJ. The amygdala: vigilance and emotion. *Mol Psychiatry* 2001;**6**:13–34

De Bellis M, Keshavan MS, Spencer S, et al. N-acetylaspartate concentration in the anterior cingulate of maltreated children and adolescents with PTSD. *Am J Psychiatry* 2000a;**157**:1175–1177

De Bellis MD, Casey BJ, Dahl RE, et al. A pilot study of amygdala volumes in pediatric generalized anxiety disorder. *Biol Psychiatry* 2000b;**48**:51–57

De Waal F. *Peacemaking among Primates*. Cambridge, MA: Harvard University Press, 1989

Delgado PL, Moreno FA. Hallucinogens, serotonin and obsessive-compulsive disorder. *J Psychoactive Drugs* 1998;**30**:359–366

Dolan RJ, Lane R, Chua P, et al. Dissociable temporal lobe activations during emotional episodic memory retrieval. *Neuroimage* 2000;**11**:203–209

Drevets WC. Neuroimaging studies of mood disorders. *Biol Psychiatry* 2000;**48**:813–829

Duchaine B, Cosmides L, Tooby J. Evolutionary psychology and the brain. *Curr Opin Neurobiol* 2001;**11**:225–230

Duman RS, Malberg J, Nakagawa S, et al. Neuronal plasticity and survival in mood disorders. *Biol Psychiatry* 2000;**48**:732–739

Dupont RL, Rice DP, Miller LS, et al. Economic costs of anxiety disorders. *Anxiety* 1996;**2**:167–172

El Mansari M, Bouchard C, Blier P. Alteration of serotonin release in the guinea pig orbito-frontal cortex by selective serotonin reuptake inhibitors. *Neuropsychopharm* 1995;**13**:117–127

Fawcett J, Stein DJ, Jobson KO. *Textbook of Treatment Algorithms in Psychopharmacology*. Chichester: John Wiley, 1999

Flint J. Freeze! *Nat Genet* 1997;**17**:250–251

Fog R, Pakkenberg H. Behavioral effects of dopamine and p-hydroxyamphetamine injected into corpus striatum of rats. *Exp Neurol* 1971;**31**:75–86

Gardner H. *The Mind's New Science: A History of the Cognitive Revolution*. New York: Basic Books, 1985

Gazzaniga MS. *The New Cognitive Neurosciences*, 2nd Edn. Cambridge, MA: MIT Press, 2000

Gelenberg AJ, Chesen CL. How fast are antidepressants? *J Clin Psychiatry* 2000; **61**:712–721

Germine M, Goddard AW, Woods SW, et al. Anger and anxiety responses to m-chlorophenylpiperazine in generalized anxiety disorder. *Biol Psychiatry* 1992;**32**:457–467

Goddard AW, Mason GF, Almai A, et al. Reductions in occipital cortex GABA levels in panic disorder detected with 1h-magnetic resonance spectroscopy. *Arch Gen Psychiatry* 2001;**58**:556–561

Goodman WK, McDougle CJ, Lawrence LP. Beyond the serotonin hypothesis: a role for dopamine in some forms of obsessive-compulsive disorder. *J Clin Psychiatry* 1990;**51S**:36–43

Goodman WK, Price LH, Rasmussen SA, et al. The Yale–Brown Obsessive Compulsive Scale. I. Development, use, and reliability. *Arch Gen Psychiatry* 1989;**46**:1006–1011

Gorman JM, Kent JM, Sullivan GM, et al. Neuroanatomical hypothesis of panic disorder, revised. *Am J Psychiatry* 2000;**157**:493–505

Grant KA, Shively CA, Nader MA, et al. Effect of social status on striatal dopamine

D2 receptor binding characteristics in cynomolgus monkeys assessed with positron emission tomography. *Synapse* 1998;**29**:80–83

Gray JA, McNaughton N. The neuropsychology of anxiety. In Hope DA (ed.) *Perspectives on Anxiety, Panic, and Fear,* Vol 43. Nebraska Symposium on Motivation. Omaha, NE: University of Nebraska Press, 1996:61–134

Greenberg LS, Safran JD. *Emotion in Psychotherapy.* New York,: Guilford Press, 1990

Greenberg PE, Sisitsky T, Kessler RC, et al. The economic burden of the anxiety disorders in the 1990s. *J Clin Psychiatry* 1999;**60**:427–435

Halligan PW, Bass C, Wade DT. New approaches to conversion hysteria. *BMJ* 2000;**320**:1488–1489

Hamner MB, Lorberbaum JP, George MS. Potential role of the anterior cingulate cortex in PTSD: review and hypothesis. *J Neuropsych Clin Neurosci* 1999;**9**:1–14

Handley SL. 5-Hydroxytryptamine pathways in anxiety and its treatment. *Pharmacol Ther* 1995;**66**:103–148

Harnad S. Other bodies, other minds: a machine incarnation of an old philosophical problem. *Minds and Machines* 1991;**1**:43–54

Heim C, Nemeroff CB. The role of childhood trauma in the neurobiology of mood and anxiety disorders: preclinical and clinical studies. *Biol Psychiatry* 2001;**49**:1023–1029

Hertzberg MA, Butterfield M, I, Feldman ME, et al. A preliminary study of lamotrigine for the treatment of posttraumatic stress disorder. *Biol Psychiatry* 1999;**45**:1226–1229

Higley JD, Mehlman PT, Taub DM, et al. Excessive mortality in young free-ranging male non-human primates with low cerebrospinal fluid 5-hydroxyindolacetic acid concentrations. *Arch Gen Psychiatry* 1996;**55**:537–543

Hollander E, Stein DJ, Broatch J, et al. A pharmacoeconomic and quality of life study of obsessive-compulsive disorder. *CNS Spectrums* 1997;**2**:16–25

Horowitz MJ. *Person Schemas and Maladaptive Interpersonal Behavior Patterns.* Chicago, IL: Chicago University Press, 1991

Hugdahl K. Cortical control of human classical conditioning: autonomic and PET data. *Pyschophysiology* 1998;**35**:170–178

Hyman SE, Nestler EJ. Initiation and adaptation: a paradigm for understanding psychotropic drug action. *Am J Psychiatry* 1996;**153**:151–162

Insel TR. A neurobiological basis of social attachment. *Biol Psychiatry* 1997; **154**:726–735

Iny LJ, Pecknold J, Suranyi-Cadotte BE, et al. Studies of a neurochemical link between depression, anxiety, and stress from [3H] imipramine and [3H] paroxetine binding on human platelets. *Biol Psychiatry* 1994;**36**:281–291

Jacobs BJ, Fornal CA. Serotonin and behavior: a general hypothesis. In Bloom FE, Kupfer DJ (eds) *Psychopharmacology: Fourth Generation of Progress.* New York: Raven Press, 1995

Johnson DL, Wiebe JS, Gold SM, et al. Cerebral blood flow and personality: a positron emission tomography study. *Am J Psychiatry* 1999;**156**:252–257

Jorm AF. Mental health literacy. Public knowledge and beliefs about mental disorders. *Br J Psychiatry* 2000;**177**:396–401

Jorm AF, Korten AE, Jacomb PA, et al. 'Mental health literacy': a survey of the public's ability to recognise mental disorders and their beliefs about the effectiveness of treatment. *Med J Aust* 1997;**166**:182–186

Kagan J, Reznick JS, Gibbons J. Biological basis of childhood shyness. *Science* 1988;**240**:167–171

Kandel ER. From metapsychology to molecular biology: explorations into the nature of anxiety. *Am J Psychiatry* 1983;**140**:1277–1293

Kendler KS, Neale MC, Kessler RC, et al. Major depression and generalized anxiety disorder: same genes, (partly) different environments? *Arch Gen Psychiatry* 1992;**49**:716–722

Kendler KS, Thornton LM, Gardner CO. Stressful life events and previous episodes in the etiology of major depression in women: An evaluation of the 'kindling' hypothesis. *Am J Psychiatry* 2000;**157**:1243–1251

Kennedy SH, Evans KR, Krüger S, et al. Changes in regional brain glucose metabolism measured with positron emission tomography after paroxetine treatment of major depression. *Am J Psychiatry* 2001;**158**:899–905

Kessler RC. The epidemiology of pure and comorbid generalized anxiety disorder. A review and evaluation of recent research. *Acta Psychiatr Scand* 2001;**406**:7–13 (review)

Kessler RC, McGonagle KC, Zhao S, et al. Lifetime and 12-month prevalence of DSM-III-R psychiatric disorders in the United States: results from the National Comorbidity Survey. *Arch Gen Psychiatry* 1994;**51**:8–19

Kessler RC, Stang P, Wittchen H-U, et al. Lifetime co-morbidities between social phobia and mood disorders in the US National Comorbidity Survey. *Psychol Med* 1999;**29**:555–567

Kessler RC, Stein MB, Berglund P. Social phobia subtypes in the National Comorbidity Study. *Am J Psychiatry* 1998;**155**:613–619

Kimbrell TA, George MS, Parekh P, I, et al. Regional brain activity during transient self-induced anxiety and anger in healthy adults. *Biol Psychiatry* 1999;**46**:454–465

Kirmayer LJ, Robbins JM, Dworkind M, et al. Somatization and the recognition of depression and anxiety in primary care. *Am J Psychiatry* 1993;**150**:734–741

Klein DF. Delineation of two drug-responsive anxiety syndromes. *Psychopharmacologia* 1964;**5**:397–408

Klein DF. Anxiety reconceptualized. In Klein DF, Rabkin J (eds) *Anxiety: New Research and Changing Concepts*. New York: Raven Press, 1981

Klein DF. False suffocation alarms, spontaneous panics, and related conditions: an integrative hypothesis. *Arch Gen Psychiatry* 1993;**50**:306–317

Kleinman A. *Rethinking Psychiatry: From Cultural Category to Personal Experience*. New York: Free Press, 1988

Klüver H, Bucy PC. Preliminary analysis of functions of the temporal lobes in monkeys. *Arch Neurol Psychiatry* 1939;**42**:979–1000

Lane RD, Reiman EM, Ahern GL, et al. Neuroanatomical correlates of happiness, sadness, and disgust. *Am J Psychiatry* 1997;**154**:926–933

Le Doux J. Fear and the brain: where have we been, and where are we going? *Biol Psychiatry* 1998;**44**:1229–1238

Leary MR, Cutlip WDI, Brit TW, et al. Social blushing. *Psychological Bull* 1992;**3**:446–460

Leckman JF, Goodman WK, North WG. The role of central oxytocin in obsessive compulsive disorder and related normal behavior. *Psychoneuroendocrinol* 1994;**19**:723–749

Leckman JF, Grice DE, Boardman J, et al. Symptoms of obsessive-compulsive disorder. *Am J Psychiatry* 1997;**154**:911–917

Lesch KP. Serotonergic gene expression and depression: implications for developing novel antidepressants. *J Affective Disord* 2001;**62**:57–76

Leuchter AF, Cook IA, Witte EA, et al. Changes in brain function of depressed subjects during treatment with placebo. *Am J Psychiatry* 2002;**159**:122–129

Liebowitz MR. Social phobia. In Ban TA, Pichot P, Poldinger W (eds) *Modern Problems of Pharmacopsychiatry*. Basel: Karger, 1987

Lightowler S, Kennet GA, Williamson IJ, et al. Anxiolytic-like effect of paroxetine in a rat social interaction test. *Pharmacol Biochem Behav* 1994;**49**:281–285

Löw K, Crestani F, Keist R, et al. Molecular and neuronal substrate for the selective attenuation of anxiety. *Science* 2001;**290**:131–134

Lucey J, V, Costa DC, Adshead G, et al. Brain blood flow in anxiety disorders. OCD, panic disorder with agoraphobia, and post-traumatic stress disorder on 99mTcHMPAO single photon emission tomography (SPET). *Br J Psychiatry* 1997;**171**:346–350

Lyons DM, Yang C, Sawyer-Glover AM, et al. Early life stress and inherited variation in monkey hippocampal volumes. *Arch Gen Psychiatry* 2001; **58**:1145–1151

MacFall JR, Payne ME, Provenzale JE, et al. Medial orbital frontal lesions in late-onset depression. *Biol Psychiatry* 2001;**49**:803–806

MacLean PD. Psychosomatic disease and the visceral brain: recent developments bearing on the Papez theory of emotion. *Psychosom Med* 1949;**11**:338–353

MacLeod AK, Byrne A. Anxiety, depression, and the anticipation of future positive and negative experience. *J Abnorm Psychol* 1993;**102**:238–247

Macmillian TM. Post-traumatic stress disorder and severe head injury. *Br J Psychiatry* 1991;**159**:431–433

Maier SF. Exposure to the stressor environment prevents the temporal dissipation of behavioral depression/learned helplessness. *Biol Psychiatry* 2001;**49**:763–773

Maier W, Gansicke A, Freyberger HJ, et al. Generalized anxiety disorder (ICD-10) in primary care from a cross-cultural perspective: a valid diagnostic entity? *Acta Psychiatr Scand* 2000;**101**:29–36

Malhi GS, Bartlett JR. Depression: a role for neurosurgery? *Br J Neurosurg* 2000;**14**:415–422

Malison RT, Price LH, Berman R, et al. Reduced brain serotonin transporter availability in major depression as measured by [^{123}I]-2β-carbomethoxy-3β-(4-iodophenyl)tropane and single photon emission computed tomography. *Biol Psychiatry* 1998;**44**:1090–1098

Mann JJ, Huang YY, Underwood MD, et al. A serotonin transporter gene promoter polymorphism (5-HTTLPR) and prefrontal cortical binding in major depression and suicide. *Arch Gen Psychiatry* 2000;**57**:729–738

Marks IM, Nesse RM. Fears and fitness: an evolutionary analysis of anxiety disorders. *Ethology and Sociobiology* 1994;**15**:247–261

Martin SD, Martin E, Rai SS, et al. Brain blood flow changes in depressed patients treated with interpersonal psychotherapy of venlafaxine hydrochloride. *Arch Gen Psychiatry* 2001;**58**:641–648

Martinot M-LP, Bragulat V, Artiges E, et al. Decreased presynaptic dopamine function in the left caudate of depressed patients with affective flattening and psychomotor retardation. *Am J Psychiatry* 2001;**158**:314–316

Mayberg HS. Frontal lobe dysfunction in secondary depression. *J Neuropsych Clin Neurosci* 1994;**6**:428–442

Mayberg HS, Brannan SK, Mahurin RK, et al. Cingulate function in depression: a potential predictor of treatment response. *Neuroreport* 1997;**8**:1057–1061

Mayberg HS, Liotti M, Branna SK, et al. Reciprocal limbic-cortical function and negative mood: converging PET findings in depression and normal sadness. *Am J Psychiatry* 1999;**156**:675–682

Mayleben M, Gariepy J, Tancer M, et al. Genetic differences in social behaviour: neurobiological mechanisms in a mouse model. *Biol Psychiatry* 1992;**31S**:216A

McDonald AW, III, Cohen JD, Stenger VA, et al. Dissociating the role of the dorsolateral prefrontal and anterior cingulate cortex in cognitive control. *Science* 2000;**288**:1835–1838

McDougle CJ, Epperson CN, Pelton GH, et al. A double-blind, placebo-controlled study of risperidone addition in serotonin reuptake inhibitor-refractory obsessive-compulsive disorder. *Arch Gen Psychiatry* 2000;**57**:794–802

McDougle CJ, Goodman WK, Leckman JF. Haloperidol addition in fluvoxamine-refractory obsessive-compulsive disorder: a double-blind placebo-controlled study in patients with and without tics. *Arch Gen Psychiatry* 1994;**51**:302–308

McGuire M, Troisi A. *Darwinian Psychiatry*. New York: Oxford University Press, 1998

McHugh PR, Slavney PR. *The Perspectives of Psychiatry*. Baltimore, MD: Johns Hopkins University Press, 1988

Mehlman PT, Higley JD, Faucher I, et al. Correlation of CSF 5-HIAA concentration with sociality and the timing of emigration in free-ranging primates. *Am J Psychiatry* 1995;**152**:907–913

Melfsen S, Osterlow J, Florin I. Deliberate emotional expressions of socially anxious children and their mothers. *J Anx Disord* 2000;**14**:249–261

Meyer JH, Swinson R, Kennedy SH, et al. Increased left posterior parietal-temporal cortex activation after D-fenfluramine in women with panic disorder. *Psychiatry Res* 2000;**98**:133–143

Mineka S, Watson D, Clark LA. Comorbidity of anxiety and unipolar mood disorders. *Annu Rev Psychol* 1998;**49**:377–412

Mishkin M, Petri H. Memories and habits: some implications for the analysis of learning and retention. In Squire LR, Butters N (eds) *Neuropsychology of Memory*. New York: Guilford Press, 1984

Montgomery SA. Obsessive compulsive disorder is not an anxiety disorder. *Int Clin Psychopharmacol* 1993;**S8**:57–62

Montgomery SA, Åsberg MA. A new depression scale designed to be sensitive to change. *Br J Psychiatry* 1979;**134**:382–389

Mundo E, Richter MA, Sam F, et al. Is the 5-HT(1Dbeta) receptor gene implicated in the pathogenesis of obsessive-compulsive disorder? *Am J Psychiatry* 2000;**157**:1160–1161

Murray CJL, Lopez AD. *Global Burden of Disease: A Comprehensive Assessment of Mortality and Morbidity from Diseases, Injuries and Risk Factors in 1990 and Projected to 2020*, Vol I. Harvard, MA: World Health Organization, 1996

Nesse RM. Is depression an adaptation? *Arch Gen Psychiatry* 2000;**57**:14–20

Nesse RM, Williams GC. *Why We Get Sick: The New Science of Darwinian Medicine*. New York: Vintage Books, 1994

Norman DA. Cognition in the head and in the world: an introduction to the special issue on situated action. *Cognitive Sci* 1993;**17**:1–6
Nutt DJ. Neurobiological mechanisms in generalized anxiety disorder. *J Clin Psychiatry* 2001;**62S**:22–27
Nutt DJ, Bell CJ, Malizia AL. Brain mechanisms of social anxiety disorder. *J Clin Psychiatry* 1998;**59**:4–11
O'Donnell D, Larocque S, Seckl JR, et al. Postnatal handling alters glucocorticoid but not mineralocorticoid messenger RNA expression in the hippocampus of adult rats. *Mol Brain Res* 1994;**26**:242–248
Ormel J, Koeter MWJ, van den Brink W, et al. Recognition, management, and course of anxiety and depression in general practice. *Arch Gen Psychiatry* 1991;**48**:700–706
Owens MJ, Nemeroff CB. Role of serotonin in the pathophysiology of depression: focus on the serotonin transporter. *Clin Chem* 1994;**40**:288–295
Pallanti S, Quercioli L, Rossi A, et al. The emergence of social phobia during clozapine treatment and its response to fluoxetine augmentation. *J Clin Psychiatry* 1999;**60**:819–823
Panskepp J. *Affective Neuroscience: The Foundations of Human and Animal Emotions.* New York: Oxford University Press, 1998
Papp LA, Klein DF, Gorman JM. Carbon dioxide hypersensitivity, hyperventilation, and panic disorder. *Am J Psychiatry* 1993;**150**:1149–1157
Pappert S. *Mindstorms: Children, Computers and Powerful Ideas.* London: Harper-Collins, 1980
Parks CL, Robinson PS, Sibille E, et al. Increased anxiety of mice lacking the serotonin 1A receptor. *Proc Natl Acad Sci USA* 1998;**95**:10734–10739
Penava SJ, Otto MW, Pollack MH, et al. Current status of pharmacotherapy for PTSD: an effect size analysis of controlled studies. *Depress Anx* 1996;**4**:240–242
Piaget J. *The Origins of Intelligence in Children.* New York: International Universities Press, 1952
Pine DS, Weese-Mayer DE, Silvestri JM, et al. Anxiety and congenital hypoventilation syndrome. *Am J Psychiatry* 1994;**151**:864–870
Pitman RK, Sanders KM, Zusman RM, et al. Pilot study of secondary prevention of posttraumatic stress disorder with propranolol. *Biol Psychiatry* 2002;**51**:189–192
Post RM. Transduction of psychosocial stress into the neurobiology of recurrent affective disorder. *Am J Psychiatry* 1992;**149**:999–1010
Potts NL, Davidson JR, Krishnan KR, et al. Magnetic resonance imaging in social phobia. *Psychiatry Res* 1994;**52**:35–42
Raleigh MJ, Brammer GL, McGuire MT. Male dominance, serotonergic systems, and the behavioral and physiological effects of drugs in vervet monkeys (*Cercopithecus aethiops sabaeus*). In Miczek KA (ed.) *Ethopharmacology: Primate Models of Neuropsychiatric Disorders.* New York: Alan R Liss, 1983
Raleigh MJ, Brammer GL, McGuire MT, et al. Dominant social status facilitates the behavioral effects of serotonergic agonist. *Brian Res* 1985;**348**:274–282
Raleigh MJ, McGuire MT, Brammer GL, et al. Social and environmental influences on blood serotonin concentrations in monkeys. *Arch Gen Psychiatry* 1984;**41**:405–410
Ramboz S, Oosting R, Amara DA, et al. Serotonin receptor 1A knockout: an

animal model of anxiety-related disorder. *Proc Natl Acad Sci USA* 1998; **95**:14476–14481

Rapoport JL, Ryland DH, Kriete M. Drug treatment of canine acral lick. *Arch Gen Psychiatry* 1992;**48**:517–521

Rauch SL, Baxter LRJ. Neuroimaging in obsessive-compulsive disorder and related disorders. In Jenicke MA, Baer L, Minichiello WE (eds) *Obsessive-Compulsive Disorders: Practical Management*, 3rd Edn. St Louis, MI: Mosby, 1998

Rauch SL, Savage CR, Alpert NM, et al. Probing striatal function in obsessive compulsive disorder: a PET study of implicit sequence learning. *J Neurosci* 1997a;**9**:568–573

Rauch SL, Savage CR, Alpert NM, et al. The functional neuroanatomy of anxiety: a study of three disorders using positron emission tomography and symptom provocation. *Biol Psychiatry* 1997b;**42**:446–452

Rauch SL, Shin LM, Whalen PJ, et al. Neuroimaging and the neuroanatomy of posttraumatic stress disorder. *CNS Spectrums* 1998;**3**:31–41

Rauch SL, van der Kolk BA, Fisler RE, et al. A symptom provocation study of posttraumatic stress disorder using positron emission tomography and script-driven imagery. *Arch Gen Psychiatry* 1996;**53**:380–387

Reber AS. *Implicit Learning and Tacit Knowledge: An Essay on the Cognitive Unconscious*. New York: Oxford University Press, 1993

Redmond DEJ. The possible role of locus coeruleus noradrenergic activity in anxiety-panic. *Clin Neuropharmacol* 1986;**9S4**:40–42

Reiman EM, Fusselman MJ, Fox PT, et al. Neuroanatomical correlates of anticipatory anxiety. *Science* 1989a;**243**:1071–1074

Reiman EM, Raichle ME, Robins E, et al. The application of positron emission tomography to the study of panic disorder. *Am J Psychiatry* 1986;**143**:469–477

Reiman EM, Raichle ME, Robins E, et al. Neuroanatomical correlates of a lactate-induced anxiety attack. *Arch Gen Psychiatry* 1989b;**46**:493–500

Richard IH, Schiffer RB, Kurlan R. Anxiety and Parkinson's disease. *J Neuropsych Clin Neurosci* 1996;**8**:383–392

Rickels K, Rynn MA. What is generalized anxiety disorder? *J Clin Psychiatry* 2001;**62S11**:4–12

Robins LN, Helzer JE, Weissman MM, et al. Lifetime prevalence of specific psychiatric disorders in three sites. *Arch Gen Psychiatry* 1984;**41**:949–958

Robins TW, Brown VJ. The role of the striatum in the mental chronometry of action: a theoretical review. *Rev Neurosci* 1990;**2**:181–213

Robins TW, Everitt BJ. Central norepinephrine neurons and behavior. In Bloom FE, Kupfer DJ (eds) *Psychopharmacology: The Fourth Generation of Progress*. New York: Raven Press, 1995

Robinson RG, Kubos KL, Starr LB, et al. Mood disorders in stroke patients: importance of location of lesion. *Brain* 1984;**107**:81–93

Rocca P, Ferrero P, Gualerzi A, et al. Peripheral-type benzodiazepine receptors in anxiety disorders. *Acta Psychiatr Scand* 1991;**84**:537–544

Rolls ET. A theory of emotion, and its application to understanding the neural basis of emotion. *Cognition and Emotion* 1990;**4**:161–190

Roy-Byrne PP, Katon W. Generalized anxiety disorder in primary care: the precursor/modifier pathway to increased health care utilization. *J Clin Psychiatry* 1997;**58S3**:34–38

Roy-Byrne PP, Stang P, Wittchen H-U, et al. Lifetime panic-depression comorbidity in the National Comorbidity Survey. *Br J Psychiatry* 2000;**176**:229–235

Saint-Cyr JA, Taylor AE, Nicholson K. Behavior and the basal ganglia. In Weiner WJ, Lang AE (eds) *Behavioral Neurology of Movement Disorders*. New York: Raven Press, 1995

Salloway S, Malloy P, Cummings JL. *The Neuropsychiatry of Limbic and Subcortical Disorders*. Washington, DC: American Psychiatric Press, 1997

Sanchez MM, Ladd CO, Plotsky PM. Early adverse experience as a developmental risk factor for later psychopathology: evidence from rodent and primate models. *Dev Psychopathol* 2001;**13**:419–449

Sapolsky RM. Glucocorticoids and hippocampal atrophy in neuropsychiatric disorders. *Arch Gen Psychiatry* 2000;**57**:925–935

Sargent PA, Kjaer KH, Bench CJ, et al. Brain serotonin $_{1A}$ receptor binding measured by positron emission tomography with [^{11}C]WAY-100635: effects of depression and antidepressant treatment. *Arch Gen Psychiatry* 2000; **57**:174–180

Schneider F, Weiss U, Kessler C, et al. Subcortical correlates of differential classical conditioning of aversive emotional reactions in social phobia. *Biol Psychiatry* 1999;**45**:863–871

Schneier FR, Johnson J, Hornig CD, et al. Social phobia: comorbidity and morbidity in an epidemiological sample. *Arch Gen Psychiatry* 1992;**49**:282–288

Schneier FR, Liebowitz MR, Abi-Dargham A. Low dopamine D2 receptor binding potential in social phobia. *Am J Psychiatry* 2001;**157**:457–459

Sedgwick P. *Psychopolitics*. London: Pluto Press, 1982

Seedat S, Stein DJ, Warwick J, et al. Single photon emission computed tomography before and after treatment with the selective serotonin reuptake inhibitor citalopram. *Int J Neuropsychopharmacol* 2000;**2S**:378

Segal ZV, Williams JM, Teasdale JD, Gemar M. A cognitive science perspective on kindling and episode sensitization in recurrent affective disorder. *Psychol Med* 1996;**26**:371–380

Serra G, Collu M, D'Aquila PS, et al. Role of the mesolimbic dopamine system in the mechanism of action of antidepressants. *Pharmacol Toxicol* 1992;**71**:72–85

Shear MK. Factors in the etiology and pathogenesis of panic disorder: revisiting the attachment separation paradigm. *Am J Psychiatry* 1996;**153**:125–135

Sheline Y, I. 3D MRI studies of neuroanatomic changes in unipolar major depression: the role of stress and medical comorbidity. *Biol Psychiatry* 2000; **48**:791–800

Sobin C, Sackheim HA. Psychomotor symptoms of depression. *Am J Psychiatry* 1997;**154**:4–17

Soubrie P. Reconciling the role of central serotonin neurones in human and animal behavior. *Behav Brain Sci* 1986;**9**:319–364

Southwick SM, Yehuda R. The interaction between pharmacotherapy and psychotherapy in the treatment of posttraumatic stress disorder. *Am J Psychotherapy* 1993;**47**:404–411

Spitzer RL, Wakefield JC. DSM-IV diagnostic criterion for clinical significance: does it help solve the false positive problem? *Am J Psychiatry* 1999; **156**:1856–1864

Stahl SM. Mechanism of action of serotonin selective reuptake inhibitors:

serotonin receptors and pathways mediate therapeutic effects and side effects. *J Affective Disord* 1998;**51**:215–235

Staley JK, Malison RT, Innis RB. Imaging of the serotonergic system: interactions of neuroanatomical and functional abnormalities of depression. *Biol Psychiatry* 1998;**44**:534–549

Steere JC, Li B-M, Jentsch JD, et al. Alpha-1 noradrenergic stimulation impairs, while alpha-2 stimulation improves, prefrontal cortical monoamine responses to psychological stress in the rat. *Soc Neurosci Abstr* 1996;**22**:1126

Stein DJ. Philosophy and the DSM-III. *Compr Psychiatry* 1991;**32**:404–415

Stein DJ. Psychoanalysis and cognitive science: contrasting models of the mind. *J Am Acad Psychoanal* 1992;**20**:543–559

Stein DJ. The neurobiology of obsessive-compulsive disorder. *Neuroscientist* 1996;**2**:300–305

Stein DJ. *Cognitive Science and the Unconscious.* Washington, DC: American Psychiatric Press, 1997

Stein DJ. Single photon emission computed tomography of the brain with Tc-99m HMPAO during sumatriptan challenge in obsessive-compulsive disorder: Investigating the functional role of the serotonin auto-receptor. *Prog Neuropsychopharm Biol Psychiatry* 1999;**23**:1079–1099

Stein DJ. The neurobiology of evil: psychiatric perspectives on perpetrators. *Ethnicity and Health* 2000;**5**:305–315

Stein DJ. Comorbidity in generalized anxiety disorder: impact and implications. *J Clin Psychiatry* 2001a;**62S**:29–36

Stein DJ. Neurobiology of the obsessive-compulsive spectrum of disorders. *Biol Psychiatry* 2001b;**47**:296–304

Stein DJ, Bouwer C. A neuro-evolutionary approach to the anxiety disorders. *J Anx Disord* 1997a;**11**:409–429

Stein DJ, Bouwer C. Blushing and social phobia: a neuroethological speculation. *Medical Hypotheses* 1997b;**49**:101–108

Stein DJ, Hollander E. Impulsive aggression and obsessive-compulsive disorder. *Psych Annals* 1993;**23**:389–395

Stein DJ, Hollander E, Cohen L. Neuropsychiatry of obsessive-compulsive disorder. In Hollander E, Zohar J, Marazziti D, Olivier B (eds) *Current Insights in Obsessive-Compulsive Disorder.* Chichester: Wiley, 1994

Stein DJ, Hollander E, Liebowitz MR. Neurobiology of impulsivity and impulse control disorders. *J Neuropsych Clin Neurosci* 1993;**5**:9–17

Stein DJ, Liu Y, Shapira NA, et al. The psychobiology of obsessive-compulsive disorder: how important is the role of disgust? *Current Psychiatry Reports* 2001;**3**:281–287

Stein DJ, O'Sullivan R, Hollander E. Neurobiology of trichotillomania. In Stein DJ, Christenson GA, Hollander E (eds) *Trichotillomania.* Washington, DC: American Psychiatric Press, 1999a

Stein DJ, Rapoport JL. Cross-cultural studies and obsessive-compulsive disorder. *CNS Spectrums* 1996;**1**:42–46

Stein DJ, Seedat S, Potocnik F. Hoarding: a review. *Israel J Psychiatry* 1999b; **36**: 35–46

Stein DJ, Shoulberg N, Helton K, et al. The neuroethological model of obsessive-compulsive disorder. *Compr Psychiatry* 1992;**33**:274–281

Stein D, Spadaccini E, Hollander E. Meta-analysis of pharmacotherapy trials for obsessive compulsive disorder. *Int Clin Psychopharmacol* 1995;**10**:11–18

Stein DJ, Stein MB, Goodwin W, et al. The selective serotonin reuptake inhibitor paroxetine is effective in more generalized and less generalized social anxiety disorder. *Psychopharmacology* 2001;**158**:267–272

Stein DJ, Stone MH. *Essential Papers on Obsessive-Compulsive Disorders*. New York: New York University Press, 1997

Stein DJ, Westenberg H, Liebowitz MR. Social anxiety disorder and generalized anxiety disorder: serotonergic and dopaminergic neurocircuitry. *J Clin Psychiatry* 2002;**63**:12–19

Stein DJ, Williams D. Cross-cultural aspects of anxiety disorders. In Stein DJ, Hollander E (eds) *Textbook of Anxiety Disorders*. Washington, DC: American Psychiatric Press, 2002

Stein DJ, Young JE. *Cognitive Science and Clinical Disorders*. San Diego, CA: Academic Press, 1992

Stein DJ, Zungu-Dirwayi N, van der Linden GJ, et al. Pharmacotherapy for post-traumatic stress disorder. *Cochrane Database of Systematic Reviews* 2000; **4**:CD002795

Stein MB, Chartier MJ, Kozak MV. Genetic linkage to the serotonin transporter protein and 5HT2A receptor genes excluded in generalized social phobia. *Psychiatry Res* 1998;**81**:283–291

Stein MB, Delaney SM, Chartier MJ. Platelet [3H]-paroxetine binding in social phobia: comparison to patients with panic disorder and healthy volunteers. *Biol Psychiatry* 1995;**37**:224–228

Stern E, Silbersweig DA, Chee K-Y, et al. A functional neuroanatomy of tics in Tourette's syndrome. *Arch Gen Psychiatry* 2000;**57**:741–748

Stevens A, Price J. *Evolutionary Psychiatry: A New Beginning*. London: Routledge, 1996

Sussman N, Stein DJ. Pharmacotherapy of generalized anxiety disorder. In Stein DJ, Hollander E (eds) *Textbook of Anxiety Disorders*. Washington, DC: American Psychiatric Press, 2002

Swedo SE, Leonard HL, Garvey M, et al. Pediatric autoimmune neuropsychiatric disorders associated with streptococcal infections: clinical description of the first 50 cases. *Am J Psychiatry* 1998;**155**:264–271

Tancer ME, Mailman RB, Stein MB, et al. Neuroendocrine sensitivity to monoaminergic system probes in generalized social phobia. *Anxiety* 1994; **1**:216–223

Tauscher J, Bagby RM, Javanmard M, et al. Inverse relationship between serotonin 5-HT_{1A} receptor binding and anxiety: a [^{11}C]WAY-100653 PET investigation in healthy volunteers. *Am J Psychiatry* 2001;**158**:1326–1328

Thase ME, Rush J, Howland RH, et al. Double-blind switch study of imipramine or sertraline treatment of antidepressant-resistant chronic depression. *Arch Gen Psychiatry* 2002;**59**:233–239

Thoren P, Asberg M, Bertilsson L. Clomipramine treatment of obsessive-compulsive disorder. II. Biochemical aspects. *Arch Gen Psychiatry* 1980;**37**:1289–1294

Tiihonen J, Kuikka J, Bergstrom K, et al. Dopamine reuptake site densities in patients with social phobia. *Am J Psychiatry* 1997a;**154**:239–242

Tiihonen JF, Kuikka J, Rasanen P, et al. Cerebral benzodiazepine receptor binding and distribution in generalized anxiety disorder: a fractal analysis. *Mol Psychiatry* 1997b;**2**:463–471

Tillfors M, Furmack T, Marteinsdottir I, et al. Cerebral blood flow in subjects with social phobia during stressful speaking tasks: a PET study. *Am J Psychiatry* 2001;**158**:1220–1226

Torgersen S. Comorbidity of major depression and anxiety disorders in twin pairs. *Am J Psychiatry* 1990;**147**:1199–1202

Tupler LA, Davidson JRT, Smith RD, et al. A repeat proton magnetic resonance spectroscopy study in social phobia. *Biol Psychiatry* 1997;**42**:419–424

Turing AM. Computing machinery and intelligence. *Mind* 1950;**59**:236

Twain M. *Following the Equator*, Vol I. New York: Harper and Brothers, 1897

Uhde T. Anxiety and growth disturbance; is there a connection? *J Clin Psychiatry* 1994;**55S**:17–27

van der Linden G, van Heerden B, Warwick J, et al. Functional brain imaging and pharmacotherapy in social phobia: single photon emission tomography before and after treatment with the selective serotonin reuptake inhibitor citalopram. *Prog Neuropsychopharm Biol Psychiatry* 1999;**24**:419–438

van der Linden GJH, Stein DJ, van Balkom AJLM. The efficacy of the selective serotonin reuptake inhibitors for social anxiety disorder (social phobia): a meta-analysis of randomized controlled trials. *Int Clin Psychopharmacol* 2000;**15S2**:15–24

van Praag HM. Inflationary tendencies in judging the yield of depression research. *Neuropsychobiol* 1998;**37**:130–141

Videbach P. PET measurements of brain glucose metabolism and blood flow in major depressive disorder: a critical review. *Acta Psychiatr Scand* 2000; **101**:11–20

Vythilingam M, Anderson ER, Goddard A, et al. Temporal lobe volume in panic disorder – a quantitative magnetic resonance imaging study. *Psychiatry Res* 2000;**99**:75–82

Wang PW, Ketter TA. Biology and recent brain imaging studies in affective psychoses. *Current Psychiatry Reports* 2000;**2**:298–304

Weissman MM, Bland RC, Canino GJ, et al. The cross national epidemiology of obsessive compulsive disorder. *J Clin Psychiatry* 1994;**55S**:5–10

Weissman MM, Bland RC, Canino GJ, et al. The cross-national epidemiology of social phobia: a preliminary report. *Int Clin Psychopharmacol* 1996;**11S**:9–14

Weizman R, Tanne Z, Granek M, et al. Peripheral benzodiazepine binding sites on platelet membranes are increased during diazepam treatment of anxious patients. *Eur J Pharmacol* 1987;**138**:289–292

Wu JC, Buchsbaum MS, Hershey TG, et al. PET in generalized anxiety disorder. *Biol Psychiatry* 1991;**29**:1181–1199

Yehuda R. *Risk Factors for Posttraumatic Stress Disorder*. Washington, DC: American Psychiatric Press, 1999

Yehuda R, McFarlane AC. Conflict between current knowledge about posttraumatic stress disorder and its original conceptual basis. *Am J Psychiatry* 1995;**152**:1705–1713

Yehuda R, Southwick SM, Krystal JH, et al. Enhanced suppression of cortisol following dexamethasone administration in posttraumatic stress disorder. *Am J Psychiatry* 1993;**150**:83–86

Young GB, Chandarana PC, Blume WT, et al. Mesial temporal lobe seizures presenting as anxiety disorders. *J Neuropsych Clin Neurosci* 1995;**7**:352–357

Zald DH, Kim SW. Anatomy and function of the orbital frontal cortex, I: Anatomy, neurocircuitry, and obsessive-compulsive disorder. *J Neuropsych Clin Neurosci* 1996;**8**:125–138

Zohar J, Insel TR, Zohar-Kadouch RC. Serotonergic responsivity in obsessive-compulsive disorder: effects of chronic clomipramine treatment. *Arch Gen Psychiatry* 1988;**45**:167–172

Appendix

Table A.1 Montgomery–Åsberg Depression Rating Scale.

1. *Apparent sadness*

Representing despondency, gloom, and despair (more than just ordinary transient low spirits) reflected in speech, facial expression, and posture. Rate depth and inability to brighten up.

0 No sadness
1
2 Looks dispirited but does brighten up without difficulty
3
4 Appears sad and unhappy most of the time
5
6 Looks miserable all the time. Extremely despondent

2. *Reported sadness*

Representing reports of depressed mood, regardless of whether it is reflected in appearance or not. Includes low spirits, despondency, or the feeling of being beyond help and without hope. Rate according to intensity, duration, and the extent to which the mood is influenced by events.

0 Occasional sadness in keeping with circumstances
1
2 Sad or low but brightens up without difficulty
3
4 Pervasive feelings of sadness and gloominess. The mood is still influenced by external circumstances

Table A.1 *continued*

5
6 Continuous or unvarying sadness, misery, or despondency

3. Inner tension

Representing feelings of ill-defined discomfort, edginess, inner turmoil, mental tension mounting to panic, dread, or anguish. Rate according to intensity, duration or extent of reassurance called for.

0 Placid. Only fleeting inner tension
1
2 Occasional feelings of edginess and ill-defined discomfort
3
4 Continuous feelings of inner tension or intermittent panic which the patient can only master with some difficulty
5
6 Unrelenting dread or anguish. Overwhelming panic

4. Reduced sleep

Representing the experience of reduced duration or depth of sleep compared to the subject's own normal pattern when well.

0 Sleep as usual
1
2 Slight difficulty dropping off to sleep or slightly reduced, light or fitful sleep
3
4 Sleep reduced or broken by at least two hours
5
6 Less than two or three hours sleep

5. Reduced appetite

Representing the feeling of a loss of appetite compared with when well. Rate by loss of desire for food or the need to force oneself to eat.

0 Normal or increased appetite
1
2 Slightly reduced appetite
3
4 No appetite, food is tasteless
5

Table A.1 *continued*

6 Needs persuasion to eat at all

6. Concentration difficulties

Representing difficulties collecting one's thoughts mounting to incapacitating lack of concentration. Rate according to intensity, frequency, and incapacity produced.

0 No difficulties in concentrating
1
2 Occasional difficulties in collecting one's thoughts
3
4 Difficulties in concentration and sustaining thought which reduces ability to read or hold a conversation
5
6 Unable to read or converse without great difficulty

7. Lassitude

Representing a difficulty getting started or slowness initiating and performing everyday activities.

0 Hardly any difficulty getting started. No sluggishness
1
2 Difficulties in starting activities
3
4 Difficulties in starting simple activities which are carried out with effort
5
6 Complete lassitude. Unable to do anything without help

8. Inability to feel

Representing the subjective experience of reduced interest in the surroundings or activities that normally give pleasure. The ability to react with adequate emotion to circumstances or people is reduced. Rate according to intensity, duration, and the extent to which the mood is influenced by events.

0 Normal interest in surroundings and in other people
1
2 Reduced ability to enjoy usual interest
3
4 Loss of interest in the surroundings. Loss of feelings for friends and acquaintances

Table A.1 *continued*

5

6 The experience of being emotionally paralysed, inability to feel anger or grief, and a complete or even painful failure to feel for close relatives or friends

9. Pessimistic thoughts

Representing thoughts of guilt, self-reproach, sinfulness, remorse, and ruin.

0 No pessimistic thoughts

1

2 Fluctuating idea of failure, self-reproach or self-deprecation

3

4 Persistent self-accusation or definite but still rational ideas of guilt or sin. Increasingly pessimistic about the future

5

6 Delusions of ruin, remorse, or unredeemable sin. Self-accusations which are absurd and unshakeable

10. Suicidal thoughts

Representing the feeling that life is not worth living, that a natural death would be more welcome, suicidal thoughts and preparations for suicide. Suicide attempts should not in themselves influence the rating.

0 Enjoys life or takes it as it comes

1

2 Weary of life. Only fleeting suicidal thoughts

3

4 Probably better off dead. Suicidal thoughts are common and suicide is considered as a possible solution, but without specific plan of action

5

6 Explicit plans for suicide when there is an opportunity. Active preparations for suicide

Table A.2 DSM-based GAD symptom severity scale (DGSS).

1. Excessive anxiety and worry

Frequency

Have you been anxious or worried? How often in the past week?

0 None of the time
1 Very little of the time (less than 10%)
2 Some of the time (approx 20–30%)
3 Much of the time (approx 50–60%)
4 Most or all of the time (more than 80%)

Intensity

How anxious or worried were you? How much distress or discomfort did this anxiety or worrying cause you this past week? How much did they interfere with your life?

0 None
1 Mild, minimal distress or disruption of activities
2 Moderate, distress clearly present but still manageable, some disruption of activities
3 Severe, considerable distress, marked disruption of activities
4 Extreme, incapacitating distress, unable to continue activities

2. Difficulty in controlling the worry

Frequency

Have you had difficulty controlling your worry? How often in the past week?

0 None of the time
1 Very little of the time (less than 10%)
2 Some of the time (approx 20–30%)
3 Much of the time (approx 50–60%)
4 Most or all of the time (more than 80%)

Intensity

Were you able to put the worries out of your mind and think about something else? (How hard did you have to try?) How much distress or discomfort was associated? How much did this difficulty in controlling the worry interfere with your life?

0 None
1 Mild, minimal distress or disruption of activities
2 Moderate, distress clearly present but still manageable, some disruption of activities
3 Severe, considerable distress, marked disruption of activities
4 Extreme, incapacitating distress, unable to continue activities

Table A.2 *continued.*

3. Restless or feeling keyed up or on edge

Frequency

Have you felt restless or keyed up or on edge? How often in the past week?

0 None of the time
1 Very little of the time (less than 10%)
2 Some of the time (approx 20–30%)
3 Much of the time (approx 50–60%)
4 Most or all of the time (more than 80%)

Intensity

How restless or keyed up or on edge were you? How much distress or discomfort did this cause you? How much did it interfere with your life?

0 None
1 Mild, minimal distress or disruption of activities
2 Moderate, distress clearly present but still manageable, some disruption of activities
3 Severe, considerable distress, marked disruption of activities
4 Extreme, incapacitating distress, unable to continue activities

4. Being easily fatigued

Frequency

Have you been easily fatigued? (What kind of things have made you felt tired?) How often in the past week?

0 None of the time
1 Very little of the time (less than 10%)
2 Some of the time (approx 20–30%)
3 Much of the time (approx 50–60%)
4 Most or all of the time (more than 80%)

Intensity

How severe has this been? How much distress or discomfort did this cause you? How much did it interfere with your life?

0 None
1 Mild, minimal distress or disruption of activities
2 Moderate, distress clearly present but still manageable, some disruption of activities
3 Severe, considerable distress, marked disruption of activities
4 Extreme, incapacitating distress, unable to continue activities

Table A.2 *continued.*

5. Difficulty concentrating or mind going blank

Frequency	*Intensity*
Have you had difficulty concentrating or your mind going blank? How often in the past week?	**How severe has this been? How much distress or discomfort did this cause you? How much did it interfere with your life?**
0 None of the time	0 None
1 Very little of the time (less than 10%)	1 Mild, minimal distress or disruption of activities
2 Some of the time (approx 20–30%)	2 Moderate, distress clearly present but still manageable, some disruption of activities
3 Much of the time (approx 50–60%)	3 Severe, considerable distress, marked disruption of activities
4 Most or all of the time (more than 80%)	4 Extreme, incapacitating distress, unable to continue activities

6. Irritability

Frequency	*Intensity*
Have you felt irritable? How often in the past week?	**How severe has this been? How much distress or discomfort did this cause you? How much did it interfere with your life?**
0 None	0 None
1 Few activities (less than 10%)	1 Mild, minimal distress or disruption of activities
2 Some activities (approx 20–30%)	2 Moderate, distress clearly present but still manageable, some disruption of activities
3 Many activities (approx 50–60%)	3 Severe, considerable distress, marked disruption of activities
4 Most or all activities (more than 80%)	4 Extreme, incapacitating distress, unable to continue activities

Table A.2 *continued.*

7. Muscle tension

Frequency

Have you experienced muscle tension? How often in the past week?

0 None of the time
1 Very little of the time (less than 10%)
2 Some of the time (approx 20–30%)
3 Much of the time (approx 50–60%)
4 Most or all of the time (more than 80%)

Intensity

How severe has this been? How much distress or discomfort did this cause you? How much did it interfere with your life?

0 None
1 Mild, minimal distress or disruption of activities
2 Moderate, distress clearly present but still manageable, some disruption of activities
3 Severe, considerable distress, marked disruption of activities
4 Extreme, incapacitating distress, unable to continue activities

8. Sleep disturbance (difficulty falling or staying asleep, or restless unsatisfying sleep)

Frequency

Have you had difficulty falling or staying asleep? How about restless or unsatisfying sleep? How often in the past week?

0 None of the time
1 Very little of the time (less than 10%)
2 Some of the time (approx 20–30%)
3 Much of the time (approx 50–60%)
4 Most or all of the time (more than 80%)

Intensity

How severe has this been? How much distress or discomfort did this cause you? How much did it interfere with your life?

0 None
1 Mild, minimal distress or disruption of activities
2 Moderate, distress clearly present but still manageable, some disruption of activities
3 Severe, considerable distress, marked disruption of activities
4 Extreme, incapacitating distress, unable to continue activities

Table A.3 Yale–Brown Obsessive-Compulsive Scale.

1. *Time occupied by obsessive thoughts*

Q: How much of your time is occupied by obsessive thoughts? How frequently do the obsessive thoughts occur?

0 = None.
1 = Mild, less than 1 hour/day or occasional intrusion.
2 = Moderate, 1 to 3 hours/day or frequent intrusion.
3 = Severe, greater than 3 and up to 8 hours/day or very frequent intrusion.
4 = Extreme, greater than 8 hours/day or near constant intrusion.

2. *Interference due to obsessive thoughts*

Q: How much do your obsessive thoughts interfere with your social or work (or role) functioning? Is there anything that you don't do because of them?

0 = None.
1 = Mild, slight interference with social or occupational activities, but overall performance not impaired.
2 = Moderate, definite interference with social or occupational performance, but still manageable.
3 = Severe, causes substantial impairment in social or occupational performance.
4 = Extreme, incapacitating.

3. *Distress associated with obsessive thoughts*

Q: How much distress do your obsessive thoughts cause you?

0 = None.
1 = Mild, not too disturbing.
2 = Moderate, disturbing, but still manageable.
3 = Severe, very disturbing.
4 = Extreme, near constant and disabling distress.

4. *Resistance against obsessions*

Q: How much of an effort do you make to resist the obsessive thoughts? How often do you try to disregard or turn your attention away from these thoughts as they enter your mind?

0 = Makes an effort to always resist, or symptoms so minimal does not need to actively resist.
1 = Tries to resist most of the time.

Table A.3 *continued.*

2 = Makes some effort to resist.
3 = Yields to all obsessions without attempting to control them, but does so with some reluctance.
4 = Completely and willingly yields to all obsessions.

5. *Degree of control over obsessive thoughts*
Q: How much control do you have over your obsessive thoughts? How successful are you in stopping or diverting your obsessive thinking? Can you dismiss them?
0 = Complete control.
1 = Much control, usually able to stop or divert obsessions with some effort and concentration.
2 = Moderate control, sometimes able to stop or divert obsessions.
3 = Little control, rarely successful in stopping or dismissing obsessions, can only divert attention with difficulty.
4 = No control, experienced as completely involuntary, rarely able to even momentarily alter obsessive thinking.

6. *Time spent performing compulsive behaviors*
Q: How much time do you spend performing compulsive behaviors? How much longer than most people does it take to complete routine activities because of your rituals? How frequently do you perform compulsions?
0 = None.
1 = Mild (spends less than 1 hour/day performing compulsions), or occasional performance of compulsive behaviors.
2 = Moderate (spends from 1 to 3 hours/day performing compulsions), or frequent performance of compulsive behaviors.
3 = Severe (spends more than 3 and up to 8 hours/day performing compulsions), or very frequent performance of compulsive behaviors.
4 = Extreme (spends more than 8 hours/day performing compulsions), or near constant performance of compulsive behaviors (too numerous to count).

7. *Interference due to compulsive behaviors*
Q: How much do your compulsive behaviors interfere with your social or work (or role) functioning? Is there anything that you don't do because of the compulsions?

Table A.3 *continued.*

0 = None.
1 = Mild, slight interference with social or occupational activities but overall performance not impaired.
2 = Moderate, definite interference with social or occupational performance, but still manageable.
3 = Severe, causes substantial impairment in social or occupational performance.
4 = Extreme, incapacitating.

8. *Distress associated with compulsive behavior*
Q: How would you feel if prevented from performing your compulsion(s)? How anxious would you become? How anxious do you get while performing compulsions until you are satisfied they are completed?
0 = None.
1 = Mild, only slightly anxious if compulsions prevented, or only slight anxiety during performance of compulsions.
2 = Moderate, reports that anxiety would mount but remain manageable if compulsions prevented, or that anxiety increases but remains manageable during performance of compulsions.
3 = Severe, prominent and very disturbing increase in anxiety if compulsions interrupted, or prominent and very disturbing increase in anxiety during performance of compulsions.
4 = Extreme, incapacitating anxiety from any intervention aimed at modifying activity, or incapacitating anxiety develops during performance of compulsions.

9. *Resistance against compulsions*
Q: How much of an effort do you make to resist the compulsions?
0 = Makes an effort to always resist, or symptoms so minimal does not need to actively resist.
1 = Tries to resist most of the time.
2 = Makes some effort to resist.
3 = Yields to almost all compulsions without attempting to control them, but does so with some reluctance.
4 = Completely and willingly yields to all compulsions.

Table A.3 *continued.*

10. *Degree of control over compulsive behavior*

Q: How strong is the drive to perform the compulsive behavior? How much control do you have over the compulsions?

0 = Complete control.

1 = Much control, experiences pressure to perform the behavior but usually able to exercise voluntary control over it.

2 = Moderate control, strong pressure to perform behavior, can control it only with difficulty.

3 = Little control, very strong drive to perform behavior, must be carried to completion, can only delay with difficulty.

4 = No control, drive to perform behavior experienced as completely involuntary and overpowering, rarely able to even momentarily delay activity.

Reproduced from Goodman et al. *Arch Gen Psychiatry* 1089;**46**:1006–1011 with permission from Wayne Goodman MD, University of Florida.

Table A.4 Panic and Agoraphobia Scale.

(A) Panic attacks

A1. Frequency

❑ 0 no panic attack in the past week
❑ 1 1 panic attack in the past week
❑ 2 2 or 3 panic attacks in the past week
❑ 3 4–6 panic attacks in the past week
❑ 4 more than 6 panic attacks in the past week

A2. Severity

❑ 0 no panic attacks
❑ 1 attacks were usually very mild
❑ 2 attacks were usually moderate
❑ 3 attacks were usually severe
❑ 4 attacks were usually extremely severe

A3. Average duration of panic attacks

❑ 0 no panic attacks
❑ 1 1–10 minutes
❑ 2 over 10–60 minutes
❑ 3 over 1–2 hours
❑ 4 over 2 hours and more

U. Were most of the attacks expected (occurring in feared situations) or unexpected (spontaneous)?

❑	**9**	**no panic attacks**		
❑ 0	❑ 1	❑ 2	❑ 3	❑ 4
mostly unexpected	more unexpected than expected	some unexpected, some expected	more expected than unexpected	mostly expected

(B) Agoraphobia, avoidance behavior

B1. Frequency of avoidance behavior

❑ 0 no avoidance (or no agoraphobia)
❑ 1 infrequent avoidance of feared situations
❑ 2 occasional avoidance of feared situations
❑ 3 frequent avoidance of feared situations
❑ 4 very frequent avoidance of feared situations

Table A.4 *continued.*

B2. Number of feared situations

How many situations are avoided or induce panic attacks or discomfort?

- ❑ 0 none (or no agoraphobia)
- ❑ 1 1 situation
- ❑ 2 2–3 situations
- ❑ 3 4–8 situations
- ❑ 4 occurred in very many different situations

B3. Importance of avoided situations

How important are the avoided situations?

- ❑ 0 unimportant (or no agoraphobia)
- ❑ 1 not very important
- ❑ 2 moderately important
- ❑ 3 very important
- ❑ 4 extremely important

(C) Anticipatory anxiety ('fear of fear')

C1. Frequency of anticipatory anxiety

- ❑ 0 no fear of having a panic attack
- ❑ 1 infrequent fear of having a panic attack
- ❑ 2 sometimes fear of having a panic attack
- ❑ 3 frequent fear of having a panic attack
- ❑ 4 fear of having a panic attack all the time

C2. How strong was this 'fear of fear'?

- ❑ 0 no
- ❑ 1 mild
- ❑ 2 moderate
- ❑ 3 marked
- ❑ 4 extreme

(D) Disability

D1. Disability in family relationships (partnership, children, etc.)

- ❑ 0 no
- ❑ 1 mild
- ❑ 2 moderate
- ❑ 3 marked
- ❑ 4 extreme

Table A.4 *continued.*

D2. Disability in social relationships and leisure time (social events like cinema, etc.)

❑ 0 no
❑ 1 mild
❑ 2 moderate
❑ 3 marked
❑ 4 extreme

D3. Disability in employment (or housework)

❑ 0 no
❑ 1 mild
❑ 2 moderate
❑ 3 marked
❑ 4 extreme

(E) Worries about health

E1. Worries about health damage

Patient was worried about suffering bodily damage due to the disorder

❑ 0 not true
❑ 1 hardly true
❑ 2 partly true
❑ 3 mostly true
❑ 4 definitely true

E2. Assumption of organic disease

Patient thought that anxiety symptoms are due to a somatic and not to a psychological disorder

❑ 0 not true, psychological disorder
❑ 1 hardly true
❑ 2 partly true
❑ 3 mostly true
❑ 4 definitely true, somatic disorder

Table A.5 TOP-8.

Which traumatic event is the most bothersome?
Event: . . .

1. Have you experienced painful images, thoughts or memories of the event which you could not get out of your mind even though you may have wanted to?
0 = not at all
1 = mild: rarely and/or not bothersome
2 = moderate: at least once a week and/or produces some distress
3 = severe: at least 4 times per week or moderately distressing
4 = extremely severe: daily or produces so much distress that patient cannot work or function socially

2. Does exposure to an event that reminds you of, or resembles the event cause you any physical response (e.g. sweating, trembling, heart racing, nausea, hyperventilating, dizziness, etc.)?
0 = not at all
1 = a little bit: infrequent or questionable
2 = somewhat: mildly distressing
3 = significant: causes much distress
4 = marked: very distressing; may have sought help because of the physical response (e.g. chest pain so severe that person was sure he or she was having a heart attack)

3. Have you avoided places, people, conversations, or activities that remind you of the event (e.g. movies, TV shows, certain places, meetings, funerals)?
0 = no avoidance
1 = mild: of doubtful significance
2 = moderate: definite avoidance of situations
3 = severe: very uncomfortable and avoidance affects life in some way
4 = extremely severe: house-bound, cannot go out to shops or restaurants, major functional restrictions

4. Have you experienced less interest (pleasure) in things that you used to enjoy?
0 = no loss of interest
1 = one or two activities less pleasurable
2 = several activities less pleasurable
3 = most activities less pleasurable
4 = almost all activities less pleasurable

Table A.5 *continued.*

5. Do you have less to do with other people than you used to? Do you feel estranged from other people?
0 = no problem
1 = feels detached/estranged, but still normal degree of contact with others
2 = sometimes avoids contact that would normally participate in
3 = definitely and usually avoids people with whom would usually associate
4 = absolutely refuses or actively avoids all social contact

6. Can you have warm feelings/feel close to others? Do you feel numb?
0 = no problem
1 = mild: of questionable significance
2 = moderate: some difficulty expressing feelings
3 = severe: definite problems expressing feelings
4 = very severe: have no feelings, feels numb most of the time

7. Do you have to stay on guard? Are you watchful? Do you feel on edge? Do you have to sit with your back to the wall?
0 = no problem
1 = mild: occasional, not disruptive
2 = moderate: causes discomfort/feels on edge or watchful in some situations
3 = severe: causes discomfort/feels on edge or watchful in most situations
4 = very severe: causes extreme discomfort and/or alters life (feels constantly on guard/socially impaired because of hypervigilance)

8. Do you startle easily? Do you have a tendency to jump? Is this a problem after unexpected noise, or if you hear or see something that reminds you of the trauma?
0 = no problem
1 = mild: occasional but not disruptive
2 = moderate: causes definite discomfort or an exaggerated startle response at least every two weeks
3 = severe: happens more than once a week
4 = extremely severe: so bad that the person cannot function at work or socially

Table A.6 Liebowitz Social Anxiety Scale.

Use the following key to answer each question:

Fear or anxiety	*Avoidance*
0 = None	0 = Never (0%)
1 = Mild	1 = Occasionally (1–33%)
2 = Moderate	2 = Often (34–67%)
3 = Severe	3 = Usually (68–100%)

		Fear or anxiety	*Avoidance*
1.	Telephoning in public		
2.	Participating in small groups		
3.	Eating in public places		
4.	Drinking with others in public places		
5.	Talking to people in authority		
6.	Acting, performing, or giving a talk in front of an audience		
7.	Going to a party		
8.	Working while being observed		
9.	Writing while being observed		
10.	Calling someone you don't know very well		
11.	Talking with people you don't know very well		
12.	Meeting strangers		
13.	Urinating in public bathrooms		
14.	Entering a room when others are already seated		
15.	Being the center of attention		
16.	Speaking up at a meeting		
17.	Taking a test		
18.	Expressing a disagreement or disapproval to people you don't know very well		
19.	Looking people you don't know very well in the eyes		
20.	Giving a report to a group		
21.	Trying to pick up someone		
22.	Returning goods to a store		
23.	Giving a party		
24.	Resisting a high-pressure salesperson		

Reproduced courtesy of Michael Liebowitz MD, Director of the Anxiety Disorders Clinic at the New York State Psychiatric Institute and Professor of Clinical Psychiatry at Columbia University.

Index